P9-CRS-497

understanding MARIJUANA

understanding MARIJUANA

a new look at the scientific evidence

Mitch Earleywine

2002

OXFORD
UNIVERSITY PRESS

Oxford New York
Auckland Bangkok Buenos Aires Cape Town Chennai
Dar es Salaam Delhi Hong Kong Istanbul Karachi Kolkata
Kuala Lumpur Madrid Melbourne Mexico City Mumbai Nairobi
São Paulo Shanghai Singapore Taipei Tokyo Toronto

and an associated company in Berlin

Published by Oxford University Press, Inc.
198 Madison Avenue, New York, New York 10016

www.oup.com

Library of Congress Cataloging-in-Publication Data
Earleywine, Mitchell.
Understanding marijuana : a new look at the scientific evidence /
Mitch Earleywine.
p. cm.
Includes bibliographical references and index.
ISBN 0-19-513893-7
1. Marijuana. 2. Cannabis. 3. Marijuana abuse. I. Title.
HV5822.M3 E27 2002
362.29'5—dc21 2001006378

9 8 7 6 5 4 3

Printed in the United States of America
on acid-free paper

Foreword

I can still recall the first time anyone ever offered me a marijuana cigarette. It happened late in the summer of 1963, when my wife Daphne and I were walking in the woods with a friend on a remote island off the coast of British Columbia. Our friend, a visiting poet from San Francisco, asked us if we had ever tried pot. When we said no, he produced a joint from his pocket and lit it with his Zippo lighter, offering us each a toke. Incensed with anger, I refused his offer and even considered reporting the incident to the local RCMP officer. I had been told many times by my uncle, then a detective on the Vancouver Vice Squad, that marijuana use was a stepping-stone to heroin use and that even a single puff could send one down the nightmare road to addiction. The visiting poet claimed that smoking marijuana stimulated the creative writing muse and that smoking pot was the current rage among artists in the Bay area. When he found out that I was an undergraduate senior majoring in psychology at the University of British Columbia, he asked me, "Do you know what the main difference is between a poet and a psychologist? The poet goes for the direct experience of life, including smoking this, while the psychologist takes notes and tries to come up with a diagnosis. I participate while you observe." I can still feel my irritation toward him to this day.

By the time I completed my Ph.D. in 1968 and joined the psychology department faculty at the University of Wisconsin the following year, marijuana use had reached an all-time high, particularly in places like Madison, where student hippies merged with protesters against the war in Vietnam and the aroma of marijuana competed with the stench of tear gas during many student demonstrations on campus. Pot was openly smoked by many faculty and graduate students alike at departmental parties. Many of my students and colleagues were interested in knowing more about the behavioral and psychological effects of marijuana, but the literature on this topic was sparse and contradictory.

After I joined the faculty at the University of Washington in the early 1970s, I planned to extend my ongoing research on cognitive and social determinants of alcohol use to the study of marijuana effects in humans. For example, I hoped to extrapolate my research on alcohol expectancies (using the balanced placebo design) to the effects of smoking either active marijuana or placebo cigarettes. I submitted a research grant proposal in response to a special, one-time RFA (Request for Applications) issued jointly by the National Institute on Alcohol Abuse and Alcoholism (NIAAA) and the National Institute on Drug Abuse (NIDA) to stimulate research comparing alcohol effects with another widely used psychoactive substance to be selected by the investigator. I proposed to study expectancy effects in both alcohol and marijuana use and to compare the effects of both drugs and their respective placebos on behavioral outcomes, ranging from tension reduction to the expression of aggression in response to psychosocial provocation. My grant application was reviewed independently by the study sections of both NIAAA and NIDA. Although my research plan received a high evaluation from reviewers at NIAAA, it failed to achieve a fundable score from NIDA, even though my proposed series of studies was identical, except for the choice of either alcohol or marijuana as the drug to be administered. When I asked my project office at NIDA why the review committee did not recommend funding for my grant, I was told (in confidence) that my proposal presented a potential "political problem" for NIDA, particularly if my findings showed marijuana to be "less of a problem" than alcohol in terms of its effects on social behaviors such as aggression. NIAAA, on the other hand, would provide funding for the alcohol research if and only if I dropped the marijuana studies. This was the first of many lessons I learned about the political controversy surrounding marijuana research in this country, a controversy that continues today in the current "War on Drugs." According to the U.S. policy of "zero-tolerance," any marijuana use in America (including use for medical purposes) is both bad and illegal.

Mitch Earleywine deserves considerable credit for his ability to provide a comprehensive and scientifically objective review of this continuously controversial topic in *Understanding Marijuana*. Readers seeking the latest and most accurate knowledge about marijuana and its effects now have a solid source of objective information to rely upon as a refreshing alternative to either the negative bias promoted by many government publications or the positive spin promoted by pro-pot publications (e.g., the magazine *High Times*). In reading the manuscript, I was continuously impressed with the breadth and depth of scholarship presented throughout the chapters. Speaking with the authority of both a noted researcher in the field of addictive behaviors and as a clinical psychologist who is experienced in behavior therapy and other treatment approaches, Dr. Earleywine provides a comprehensive review of the marijuana literature, ranging from basic biological mechanisms of action to promising approaches to the prevention and treatment of marijuana problems. The range of topics covered is indeed impressive. Readers are provided with a vast array of information about this amazing "green weed," from its historical origins dating back to 8000 B.C. to the present day.

In chapter 1, we learn about the important role hemp products have played over time in the development of both fabric (e.g., rope) and paper products. In chapter 2, we learn about past and current patterns of marijuana use, including the fact that up to one-third of Americans reported smoking marijuana at least once, and that fully 5% reported use in the past month. Next, we discover that the "stepping-stone" theory first told to me by my uncle in Vancouver is not supported by contemporary research—the vast majority of those who have tried marijuana do not in fact go on to harder drugs such as heroin. In chapters 4 and 5, we learn about the latest research findings about marijuana's effects on thought and memory and on its subjective effects such as emotional mood state. Chapter 6 is a clear and succinct description of the pharmacological action and effects of marijuana ingestion followed in chapter 7 by a discussion of its associated effects on mental and physical health; as noted by the author, "These results confirm that marijuana is neither completely harmless nor tragically toxic."

The timely topic of medical marijuana is addressed in chapter 8. Given that several states, mostly on the West Coast (including my home state of Washington), have passed public referendums supporting the use of marijuana for the treatment of several medical disorders (e.g., reduction of nausea associated with chemotherapy, appetite enhancement in the treatment of AIDS, lowering intraocular pressure in the treatment of glaucoma, and so on), the public need for accurate information on this topic has increased enormously. Critics of marijuana use often downplay its potential medical benefits, since to admit this would threaten the zero-tolerance doctrine that any use of this plant is both bad and illegal. "There is no firm research finding to support the proposed medical benefits of marijuana," goes the official government response to this question. The catch-22 here is that the government, including NIDA, has consistently failed to approve almost any research proposal designed to evaluate the potential medical benefits, except for a single study currently being conducted at the University of California in San Francisco to assess marijuana's impact on appetite enhancement in patients treated for AIDS.

Social problems often attributed to marijuana use, including the "amotivational syndrome," impaired driving, and aggression, are covered in chapter 9. In terms of aggressive responding, the author concludes that "the drug's absence of an impact on hostility has led every major commission report to conclude that cannabis does not increase aggression." Compared to alcohol, which clearly increases the risk of aggressive responding, marijuana has the opposite effect. No wonder my research proposal comparing alcohol and marijuana's impact on aggression was not funded by NIDA.

The topics of marijuana law and policy are well delineated in chapter 10, including the pros and cons of both the current prohibition policy and the alternative approaches of either decriminalization or legalization. The policy arguments are debated in terms of both their moral and pragmatic roots. While some countries have adopted a decriminalization policy (the Netherlands) and others are actively considering this option (Switzerland and Canada, including my home province of British Columbia), the current U.S. policy prohibiting any lawful marijuana use is unlikely to change in the near future.

Treatment for marijuana problems is covered in chapter 11. Various treatment modalities are addressed, including cognitive-behavioral therapy, twelve-step facilitation, and motivational interviewing. The material presented here shows that people who are motivated to eliminate their use of marijuana (abstinence goal) or to minimize its negative effects (harm reduction goal) have several promising alternative treatment approaches to choose from.

In his final thoughts in the last chapter, Dr. Earleywine cautions readers that due to the complexity and political context of marijuana research, "any attempt to explain this research may say more about the explainer than the explained." True, but in this case I strongly endorse his even-handed efforts in presenting relevant research and related issues about marijuana in a balanced and objective fashion in a way that will appeal to the interested reader. He also shows a keen sense of humor in getting his points across. In the final pages, he urges compassion in our response to those who choose to use marijuana, despite its continuing status as an illegal drug. As he states in his final sentence, "Perhaps we could tolerate people who want to use marijuana without causing harm to themselves or others."

G. *Alan Marlatt*
University of Washington

Preface

As I answer questions about marijuana, I am constantly reminded of three ideas. The first idea is Einstein's oft-quoted expression that everything should be made as simple as possible, but no simpler. The research on marijuana is extensive, lending itself to incomplete summary and interpretation. A description of marijuana can stress parallels to medicine, harmless intoxicants, or addictive narcotics. Perhaps all of these are true, in part, but they remain incomplete. A narrow focus prevents these descriptions from providing a full picture of the substance. These attempts at quick summaries may end up presenting cannabis as simpler than it is. I have tried to avoid this problem by providing a thorough look at all of the available research, no matter how confusing or contradictory.

The second idea concerns the distinction between research and its meaning. Data are separate from interpretations. Investigators present data and then interpret results. Sometimes results support an author's conclusions. Sometimes a close look reveals that the conclusions are unjustified. The interpretations often remind me of responses to those splotchy blots of ink from the infamous Rorschach test. People purportedly see these ambiguous pictures in a way that reveals more about them than the ink.

Authors may respond the same way to marijuana research. Their interpretations may tell more about their own biases than the data. For example, prohibitionists might mention that THC often appears in the blood of people involved in auto accidents. Yet they might omit the fact that most of these people also drank alcohol (see chapter 9). Antiprohibitionists might cite a large study that showed no sign of memory problems in chronic marijuana smokers. Yet they might neglect to mention that the tests were so easy that even a demented person could perform them (see chapter 4). I have tried to avoid this problem by providing appropriate detail about research, so that readers can interpret results for themselves.

The last idea may be the most controversial: some things are neither good nor evil. The common human desire to split the world into two categories is understandable. Decisions are easier when everything is black and white. Yet the world remains in frustrating but glorious color. Forcing everything into two categories can be a depressing and futile task. Every year fire warms some people and kills others. Water quenches thirst but also drowns. Aspirin relieves pain or causes overdose. Labeling these as good or evil requires many caveats and may be a pointless task. Perhaps marijuana is the same, but no one can decide without the appropriate information. Here's a presentation of a vast literature for those who prefer to think for themselves than be told what to think.

Acknowledgments

I'm to blame for all that's in here, but plenty of folks helped immensely. My hearty thanks to all of the following, who supported this book in many ways: Catharine Carlin, Alexandria Cymrot, Bob Earleywine, Dahlia Earleywine, Joe Earleywine, Tamar Gollan and Mark Myers, David and Felice Gordis, Lester Grinspoon, Josh Gross, Evan Helmuth, Sara Hopfauf, Jack Huntington, Leslie Iversen, Susan Kesser, Joe LaBrie, Boaz Levy, Linda Mathews, Norman Miller and Vicki Pollock, Kristine Rapan, Lucy Ravitch, David Rothblum, Aaron Saiger, Jim De Santis, Domenico Scarlatti, Jason Schiffman, John Storr, Susan and Clark Van Scoyk, Brian Varnum, and Terry Wager. Special thanks to the Wednesday Night Group, my Psychology 680 seminar, and all the students who took Psych 165, "The Drug Class!"

My dear wife, Elana, read every word of this book and many that did not make it. She diagnosed and endured my monthly bouts of PMS (pot manuscript syndrome) and went above and beyond every clichéd expression for tolerance, perseverance, and understanding. I dedicate this work to her.

Contents

understanding MARIJUANA

1

Highlights in the History of Cannabis

Humans use nearly every part of the infamous green weed *Cannabis sativa*. The plant grows quickly in many environments and can reach a height of 20 feet. Few natural pests attack the crop; few extremes in weather challenge its growth. The leaves consist of five or more narrow leaflets, each radiating from a slender stem attached to a thick, hollow stalk. The jagged edge of each leaflet resembles the blade of a serrated knife. The species is dioecious, meaning both female and male varieties of the plant exist. The male grows taller, topped by flowers covered with pollen. The shorter female plant, with its larger, pollen-catching flowers, produces seeds and protects them with a sticky resin. The stalks help produce fiber; the seeds provide food and oil. The flowers, leaves, and resin appear in medical and intoxicating preparations. Each day, smiling teens buy hemp shirts. Retailers sell snacks made from the seed. Glaucoma patients puff cannabis cigarettes in hope of saving their sight, and many people worldwide inhale marijuana smoke in an effort to alter consciousness. These industrial, medical, and recreational uses for the plant go back thousands of years, contributing to its spread from Taiwan throughout the world. Although industrial hemp, medical marijuana, and cannabis the intoxicant stem from the same species, in many ways they each have their own histories.

A History of Industrial Hemp

Unlike most plants that provide drugs, hemp provides dozens of products. None of these items contains meaningful amounts of tetrahydrocannabinol (THC), the main psychoactive ingredient in marijuana. In contrast to psychoactive cannabis plants, which contain 2% THC or more, industrial hemp often contains as little as .15% THC (Kirk, 1999). At this concentration, smoking a whole field would not create intoxication. Hemp provides fiber, cloth, paper, and food, as well as many soaps, shampoos, and oils. People grew hemp widely for these industrial uses, which helped the plant spread from Asia, through India, Africa, Europe, and the Americas.

Hemp Fiber

Hemp fibers were likely the first in history. Archeologists in Taiwan uncovered strands decorating clay pots from 8000 B.C. (Chang, 1968; Kung, 1959). Comprehending events from so long ago can prove difficult. This first use of cannabis fibers precedes recorded history, in an era when the world was markedly different. Humans may have barely understood how to cultivate plants. Other fibers were not available. Linen production did not begin until 3500 B.C. Another millennium passed before cotton's cultivation in 2500 B.C. (Grun, 1982). Hemp preceded each by thousands of years. No one knows the genius who first turned cannabis stalks into strands, or who discovered that twisting many together added strength, but these innovations started a long and productive career for the plant. New uses for hemp developed throughout the last 10,000 years. The plant's medicinal and intoxicating properties increased in popularity, too, but not until much later.

One early use of hemp demonstrates its potential danger. Ancient Chinese archers used the plant to make bowstrings. These new bowstrings proved superior to those made of bamboo. They fired arrows farther and more forcefully. Although no one has ever died from an overdose of smoked marijuana (Petro, 1997a), the plant may have killed many by helping sling sharp arrows into tender flesh. Bowstrings, however, were less important than rope, one of hemp fiber's most important uses for thousands of years.

The need for strong and durable rope, particularly in sailing, helped hemp become popular throughout the world. One of the first leaders in the manufacture of rope was also an early cultivator of hemp: southern Russia. Their first harvests began in the seventh century B.C. (Rubin, 1975). Around 200 B.C., Hieron II of ancient Greece transported hemp from France to make ropes for his ships (Stefanis, Ballas, & Madianou,

1975). Pliny the Elder (23–79 A.D.) recorded hemp's use in creating rope in his book *The Natural History*. Hemp ropes found in England date from 100 A.D. The plant for these cords was obviously imported, for the area did not cultivate cannabis until 400 A.D. Vikings hauled hemp ropes with them to Iceland and took hemp seeds on their travels as early as 850 A.D. (Abel, 1980).

Because these ropes contributed to the success of ocean voyages, hemp cultivation and production grew more important. By the end of the first millennium A.D., Italian ships outfitted with hemp rope dominated the oceans. Venice grew particularly famous for its fibers and ropes, which had a reputation for quality. Venetian hemp spinners belonged to a special guild. The laws governing their work were very strict; penalties for poor products included harsh fines and beatings. Ships built in Venice required the best riggings made from only the highest quality hemp. By the 1500s, England's developing navy required more rope than ever. In 1533, King Henry VIII decreed that every farmer should raise some hemp. Those who refused paid a fine. Henry's daughter, Queen Elizabeth I, later increased the penalty (Abel, 1980), but farmers preferred the fine to raising hemp. Despite growing demand, the plant brought little income. Its production had other drawbacks, too. The crop had to be placed aside to allow the resin to weaken and release the fibers from the stalk. These rotting plants stunk, making whole farms smell bad (Tusser, 1580). Birds who ate the hemp seeds purportedly tasted peculiar, too (Thistle & Cook, 1972).

Because the local farmers were unwilling to raise hemp, England imported it. By the early 1600s, most of England's hemp came from Russia. Unfortunately, the Russians of the era had a reputation for unsavory business practices. Since hemp sold by weight, some Russian exporters allegedly threw trash into the hemp bales to increase the weight and price. In an effort to decrease dependence on Russia, England commanded the American colonies to grow the crop (Abel, 1980). Residents of the New World hardly helped in the quest for more hemp. Jamestown colonists grew quality plants by 1616 (Gray, 1958), but they devoted larger plots to a more lucrative and deadly crop—tobacco (Bishop, 1966). By 1629, shipbuilding started in Salem, Massachusetts. Merchants there purchased every stalk of hemp available. New England towns built their own rope factories known as ropewalks. Irish immigrants adept at spinning hemp into thread opened American spinning schools. By the 1700s, the colonies used all their locally grown hemp. Ben Franklin himself told

England that the colonists did not have enough hemp to meet their own needs, so they could hardly export it (Carrier, 1962).

Hemp production altered America in many ways. By the time of the revolutionary war, Kentucky began cultivation. Their crop, however, was not a favorite among Americans. Northern manufacturers preferred imported hemp's superior quality. Henry Clay, the legendary "great compromiser" and congressman from the state, helped promote Kentucky's products. He encouraged Congress to pass high tariffs on imported hemp, increasing its price (Eaton, 1966). Production increased the slave trade enormously, because harvesting and processing the plant was arduous, time-consuming work. The Civil War interfered with this production dramatically. With no northern merchants buying the crop, southern farmers had little reason to grow it. The sale of hemp products decreased after the war, in part, because the technology for processing cotton improved. Nevertheless, hemp did not disappear. In 1890, the United States still manufactured hemp ropes. Demand also increased during both world wars. During World War II, the government even used the classic, patriotic film *Hemp for Victory* (1942) to encourage farmers to grow the crop. Subsequent laws against the plant decreased the production of hemp rope markedly since those times. Most ropes today are made of nylon or cotton fiber. Nevertheless, products of hemp fiber remain available. Hammocks fashioned from the ropes still sell today. Other old and new products require weaving the fiber into fabric.

Hemp Fabric

Humans have fashioned cloth from hemp for at least 2,400 years, but the fabric has varied dramatically in popularity. Interlacing hemp fibers to form cloth helped minimize the need for animal skins. Those of us with central heat may find it difficult to imagine the importance of hemp fabric for staying warm. These clothes may have also felt more comfortable than wearing a piece of some dead beast. This fabric became popular in many parts of the ancient world. Herodotus's *Histories* from 450 B.C. mentioned hemp clothing in the land of the Scythians and ancient Greeks. The Chinese *Book of Rites* (from around 200 B.C.) discussed hemp fabric (Rubin, 1975), and a sample of hemp cloth from the Chou dynasty (1122–249 B.C.) was uncovered in 1972 (Li, 1974). The ancient Chinese even used the plant to make shoes (Abel, 1980). The fabric remained popular in the Middle Ages. The ornate tomb of the French queen Arnegunde, who died in 570 A.D., contained gold, jewels, and a hemp cloth (Werner, 1964). Some suggest that including hemp cloth in the tomb reflects the fabric's regal stature (Abel, 1980).

Hemp cloth's prestigious position eventually faded when cotton grew more popular in the 1800s. As Europeans invaded the Americas, they initially used tons of hemp for fabric. The cold winters required warm clothes, and hemp provided the necessary cloth. But demand for hemp fabric decreased by the end of the Civil War, just as it had for the fiber. Cotton's improved harvesting and processing diminished hemp's importance (Bonnie & Whitebread, 1974). Synthetic fabrics also contributed to hemp's reduction. Current ecological concerns may help facilitate hemp's return. Modern hemp farmers report that cotton yields less fiber per acre, requires more water and pesticides, and may cause more harm to the environment. In addition, hemp may also have less negative impact on water and land than synthetic fibers made from petrochemicals.

Woven hemp products have regained some of their old popularity. Hats, shirts, pants, and even wedding gowns (Evans, 1999) employ the ancient fabric. Many hemp companies advertise as ecologically friendly and economical sources of cloth. Larger, mainstream companies are also experimenting with the manufacture and marketing of hats and shirts made from hemp. Because the plant remains illegal in the United States, American manufacturers must import their raw materials. Soon, however, American-grown hemp may return. Kentucky police arrested actor Woody Harrelson for planting four hemp seeds in protest of Kentucky's laws against the crop (Rosenthal & Kubby, 1996). Legislation passed in Hawaii and North Dakota in 1999 may permit the first American hemp harvests in over 50 years, provided the Drug Enforcement Agency (DEA) agrees to issue appropriate licenses.

Hemp Paper

For hundreds of years, ancient authors wrote on stone, wood, bamboo, silk, or parchment (Garraty & Gay, 1981; Grun, 1982). Records reveal that hemp paper initially appeared around the first century B.C. Despite this evidence, legends attribute the invention to the Chinese bureaucrat Ts'ai Lun, a creative and flamboyant character, who purportedly unveiled his new product in 105 A.D. Ts'ai Lun's dramatic story may help explain why he gets credit for the initial creation of paper. The inventor's contemporaries showed little enthusiasm for his innovative writing tool. In an effort to increase paper's popularity, Ts'ai Lun claimed the substance could raise the dead—a remarkable feat for any new product. He feigned his own death and had people burn paper to ostensibly bring him back to life (Carter, 1968). The tradition of burning paper at funerals spread through many parts of Asia. Ts'ai Lun not only marketed his invention with this dramatic display, but he also created a new use for it. Other people began burning paper at funerals, too. Soon this inexpensive and lightweight material replaced the silk and bamboo used for writing.

The technique for manufacturing paper required two of China's native plants: hemp and the mulberry tree. The bark of the tree and the fiber of the plant were ground together to form a pulp. Manufacturers then covered the pulp with water. Intermingled fibers eventually rose to the top and were placed in molds to dry as paper. The process remained a Chinese secret for hundreds of years, but it eventually spread to Japan. The Arabs learned the process in the 900s A.D., perhaps from Chinese prisoners captured during the Battle of Samarkand. Spain and other countries subsequently mastered the procedure, and within a few hundred years paper mills appeared throughout Europe (Abel, 1980).

Most paper manufacturers today use wood pulp. Hemp, however, may prove more economical and ecologically appropriate than wood. The plant is ready to harvest more quickly than trees, and it causes less damage to the ecosystem (Kirk, 1999). Hemp's illegal status prevents large-scale experiments with its use in paper production in the United States. Importing adds to the expense of raw materials, making hemp paper more expensive than some paper fashioned from wood pulp. In fact, some assert that the Marijuana Tax Act of 1937, which placed a prohibitive tax on hemp, arose when the owners of forests and chemical companies decided to eliminate alternative sources of paper (Herer, 1999; Rosenthal & Kubby, 1996). Despite its slightly higher price, the paper remains available today.

Hemp Food

The first use of hemp as food remains unknown, but examples of dishes made from the edible plant appear all across the world. Galen (130–200 A.D.) mentions that wealthy Romans enjoyed an elaborate hemp-seed dessert (Abel, 1980). For many years in Poland and Lithuania, cooks served a hemp-seed porridge known as *semieniatka* on Christmas Eve

(Benet, 1975). Natives of India report that hemp remains the favorite food of the god Shiva, as it has been for thousands of years.

Hemp seeds have also served as a food for nutritional reasons. Once they are cleaned they possess little THC, but they do contain 20 to 25% protein, including all the amino acids essential for human health. The seeds also contain many necessary minerals, such as calcium and potassium. Yet they have few of the toxic heavy metals like strontium, mercury, or arsenic. The seeds are roasted and ground into flour or pressed to derive oil. The flour retains the protein and minerals. The oil from the seeds contains essential fatty acids, which humans require to digest certain vitamins and build new cells (Wirtshafter, 1997). Hemp contributes to many foods today, even in the United States, where growing the plant is illegal. Roasted hemp seeds are a popular snack. They are pressed with nuts and honey into health-food bars. One company grinds them with peanuts to make a buttery spread. Another uses the oil in a salad dressing. Others bake the flour into chips and pretzels that contain extra protein. *The Hemp Seed Cookbook* contains more than 20 relevant recipes (Miller & Wirtshafter, 1991). Even animals eat the plant. Bird feed continued to contain sterilized seeds, even after the first federal legislation against cannabis appeared in the Marijuana Tax Act of 1937.

A History of Medical Marijuana

The cannabis plant's history as the source of hemp is separate from its story in medicine. Cannabis's use as a treatment for a variety of illnesses helped it spread from ancient Asia throughout the world. The plant consistently appeared in pharmacopoeia and folk medicine as a treatment for pain, seizure, muscle spasm, poor appetite, nausea, insomnia, asthma, and depression. Its potential to alleviate labor pains, premenstrual symptoms, and menstrual cramps also received attention in multiple medical reports from ancient times to the present. Marijuana's possible medical application has continued to increase its popularity, even with individuals who would frown upon recreational use. Therapeutic cannabis has also provided intriguing scientific and legal research, as discussed later in this book. The history of marijuana as medicine is extensive and includes many characters on many continents.

Medicinal use of cannabis began around 2737 B.C., long after its first use as a fiber. The mystical Chinese emperor Shen Neng introduced these pharmaceutical uses of cannabis. He also discovered many other medicines. These included ephedra, a natural stimulant that helped asthma and led to the invention of amphetamine; camellia sinensis, the first caffeinated tea; and ginseng, the popular herbal panacea (Aldrich, 1997). Legends often develop around people who make new discoveries, and Shen Neng is no exception. He purportedly could see into his own stomach and observe the impact of herbs on his body, making him a phenomenal authority on their pharmaceutical effects (Wallnofer & Von Rottauscher, 1965). Emperor Shen Neng prescribed cannabis tea for gout, malaria, beriberi, rheumatism, and, curiously, poor memory (Abel, 1980). Although other treatments have developed for most of these maladies, marijuana's impact on rheumatism remains part of modern research (Turner & ElSohly, 1981).

The world and its medicines were markedly different 5,000 years ago. For example, Egyptians introduced the 365-day calendar year about this time. A Chinese court musician fashioned the first bamboo flute, and the Cheops pyramid was still relatively new (Grun, 1982). Medicine remained akin to magic, so some treatments were not ideal. Shen Neng's herbs probably provided as much relief as any alternatives in the era, particularly in China. A few other drugs were available in other parts of the world. Opium was popular in the Middle East but had not reached China yet (Blum, 1984; Scott, 1969). Natives of South America chewed coca leaves roughly 200 years after medicinal marijuana's discovery (2500 B.C.), but they lived too far away to provide pharmaceuticals to the East (Maisto, Galizio, & Connors, 1995). Thus, marijuana was probably one of the best medicines available for many ailments. The Chinese understood the intoxicating side effects of this popular remedy. Physicians warned against large doses, which led to "seeing devils" and "communicating with spirits" (Li, 1974, 1975). Nevertheless, medicinal use of the plant continued, just as the use of other psychoactive drugs progressed.

Cannabis eventually spread from China to India. By 1400 B.C., the sacred Indian text *Atharvaveda* listed marijuana as a holy plant that could relieve stress. Given Hindu sanctions against the consumption of alcohol, cannabis remained one of the few substances appropriate for alleviating anxiety in this culture. The plant's notorious drying of mucous membranes, such as the "cotton mouth" reported by contemporary users, led the ancient Indian healer Sushruta to prescribe it for congestion. Sushruta also recommended the drug for fevers or inflammation of the mucous membranes (Aldrich, 1997). Other healers of India prescribed it successfully for coughs and asthma, and unsuccessfully for leprosy and dandruff (Nahas, 1990).

Medical marijuana spread farther while new uses for it developed back in China. In ancient Rome, Pliny the Elder mentioned marijuana's use as a painkilling analgesic but warned that excessive consumption could cause impotence. The Romans learned of cannabis's pharmaceutical properties as they stormed through new countries. A physician in Nero's army named Pedacius Dioscorides recommended the juice of the marijuana seed for earaches. (Later research confirmed the efficacy of this treatment [Kabilek, Krejci, & Santavy, 1960].) In 70 A.D., Dioscorides compiled a pharmacopoeia listing marijuana among the many exotic plants with medical applications. Galen, the ancient Greek doctor whose impact on Western medicine lasted centuries, used the drug to treat pain and flatulence. Back in China, Shen Neng's teachings remained well known. Around 200 A.D., the first pharmacopoeia of the East, based on his work, listed marijuana as a medicine (Abel, 1980). The ancient Chinese founder of surgery, Hua T'o, used cannabis combined with alcohol as an anesthetic around this time (Li, 1974, 1975).

Evidence suggests that new uses of the drug developed outside of China, Greece, and Rome. One novel application concerned the labor of childbirth. Marijuana traces appeared in the archeological remains of a young girl from the fourth century A.D. She apparently died in Jerusalem while giving birth (Zias et al., 1993). The medicine may have eased pain and increased uterine contractions (Aldrich, 1997). This obstetric use continued at least into the 1800s (Grigor, 1852). Women in Cambodia and Vietnam ingest tea made from marijuana to alleviate postpartum distress even today. Studies of fetal exposure to marijuana have produced mixed results and considerable controversy (Dreher, 1997), but the practice of using cannabis during delivery apparently began at least 2,400 years ago.

In addition to many reports of medicinal effects, warnings against abuse also continued. As happens today, some ancient concerns about the negative side effects of cannabis arose from confusing reports of its consequences. By 1000 A.D., some of the first concerns of the drug's ill effects appeared in Ibn Wahshiyah's Arabic text *On Poisons*. Wahshiyah warned that hashish (a potent cannabis intoxicant similar to charas) renders one blind and mute, eventually leading to continuous wretching and death (Levey, 1966). In fact, no documented cases of fatal overdose exist

(Petro, 1997a). Such cautions may have been more extreme in the Arab world because hashish remained more popular there. People in the Arab countries used more of the drug, increasing concern about abuse. Fewer warnings appeared in Europe at this time, perhaps because cannabis intoxicants remained extremely uncommon. Medical marijuana appeared as an ingredient in a popular European ointment of the era (Grattan & Singer, 1952), but warnings about the drug were rare.

By the twelfth century, marijuana had reached from Egypt to the rest of Africa. Archeologists in Ethiopia uncovered pipes containing traces of cannabis from the 1300s (Van der Merwe, 1975). Du Toit (1975) suggests that the Bantus may have brought the plant down the eastern coast of Africa. Dagga, as cannabis was known in Africa, had a medical reputation that varied from tribe to tribe. Hottentots and Mfengu prescribed it for snakebites. The Sotho used cannabis during childbirth as the natives of Jerusalem had done. Residents of Rhodesia also used marijuana to treat anthrax, dysentery, and malaria. In South Africa, the drug served as an asthma treatment. Although recreational use certainly contributed to the spread of marijuana from Egypt to the rest of Africa, well-known medicinal uses continued and new ones developed (Du Toit, 1980).

Increased travel to the Middle East, Africa, and India invariably led to more European publications addressing medical marijuana. Francois Rabelais, the French doctor and humorist, published his famous book *Gargantua and Pantagruel* in 1532. The text describes Pantagruelion (cannabis), which he claimed would ease the pains of gout, cure horses of colic, and treat burns. Garcia da Orta, a Portuguese physician who lived in the Indian city of Goa, described many herbal remedies, including marijuana's ability to enhance appetite (da Orta, 1563). By 1578 in China, Li Shih-Chen wrote of cannabis's antiemetic and antibiotic effects. The medical reputation of marijuana continued into the seventeenth century. In 1621, Robert Burton's *Anatomy of Melancholy* suggested medical marijuana may aid mood disorders. This novel idea guided research on and off, even as late as the 1970s (Grinspoon, 1971; Grinspoon & Bakalar, 1997).

Hemp received its scientific name in the latter half of the eighteenth century. In 1753, Linnaeus, the Swedish naturalist who classified nearly every living thing, dubbed the plant *Cannabis sativa*. He placed the species in the small family known as *Cannabinaceae*, which includes only cannabis and the hop plant, *Humulus lupulus*. In 1783, Lamarck, the man whose infamy stems from his incorrect hypotheses about evolution, sought to distinguish the hemp plants of Europe from those in India. He suggested a separate species native to India, *Cannabis Indica*, known for its shorter stature and greater quantity of resin. Much later, in 1924, a team of Russian botanists identified *Cannabis ruderalis*, a third species shorter than the other two. Whether all these types are variations on one plant or serve as separate species remains hotly debated even today (Schultes, Klein, Plowman, & Lockwood, 1975).

At the same time as the initial debates on classification, marijuana's medical reputation spread to the Americas. The 1764 edition of *The New England Dispensatory* recommended hemp roots to treat inflamed skin. The 1794 *Edinburgh New Dispensary* prescribed marijuana oil for many problems, including incontinence, coughs, and venereal disease. Nevertheless, few medical professionals in Europe or America prescribed the drug often. Its lack of popularity may stem from several factors. First, the treatments may not have been optimal. For example, marijuana oil likely did little to help syphilis. In addition, the potency of available cannabis extracts and tinctures varied considerably. But as more case studies attested to the drug's utility, the demand for it increased. Financial incentives grew with this demand, and soon quality preparations became more accessible. One man often receives credit (or blame) for this increased interest in medicinal cannabis in England and the Americas, the Irish physician William O'Shaughnessy.

In 1833, O'Shaughnessy worked for the British East India Company and the Medical College of Calcutta. Marijuana already served as a common remedy in India, inspiring O'Shaughnessy to investigate the drug's impact on many maladies. His results appeared in 1842 in the journal *Transactions of the Medical and Physical Society of Bombay*. O'Shaughnessy's first experiments used animals; he noted greater intoxication in carnivorous species. Whether or not carnivorous humans report greater sensitivity to marijuana compared to vegetarians remains unknown. These animal studies minimized O'Shaughnessy's fears of adverse effects from the drug, inspiring him to perform research on humans. He first administered marijuana to people with rheumatism. The treatment eased their pain, just as Shen Neng had suggested a few millennia earlier. Patients also reported enhanced mood and appetite. O'Shaughnessy's attempts at alleviating the discomfort associated with rabies, cholera, tetanus, and epilepsy also met with some limited success. Although medical marijuana did not cure these diseases, it eased the pain, nausea, and spasticity that often accompanies them. As word of this work reached England and the Americas, the drug grew increasingly more popular (Abel, 1980).

The use of marijuana also increased in France around this same time. Dr. Jacques Joseph Moreau de Tours began his experiments with cannabis in the late 1830s. His primary work focused on the cognitive effects of the drug, particularly on the parallels between intoxication and mental illness. (This research and its connection to the Hashish Club appear in the discussion of cannabis as an intoxicant in the next section.) Moreau also investigated marijuana as a treatment for depression, as Robert Burton had suggested in 1621. His results were not encouraging (Moreau, 1845). Modern research on THC's utility as an antidepressant also has produced mixed results, sometimes confirming Moreau's initial reports, sometimes helping specific individuals in certain cases (Grinspoon, 1971; Grinspoon & Bakalar, 1997).

Back in the Americas, the Ohio State Medical Society met in 1860 to summarize the medical uses of marijuana. The conference reported favorable outcomes for treating pain, inflammation, and cough (McMeens, 1860). The 1868 *U.S. Dispensatory* listed pages of medical uses for tincture of cannabis, an extract often formed by soaking marijuana in alcohol. The extract purportedly improved appetite, sexual interest, mental disorders, gout, cholera, hydrophobia, and insomnia (Wood & Bache, 1868). The drug's medical reputation had also continued in England. In 1890, Sir J. Russell Reynolds, chief physician to Queen Victoria, praised the drug in the prestigious medical journal *Lancet*. He claimed cannabis successfully treated insomnia, facial tics, asthma, and menstrual problems (Reynolds, 1890). The queen herself allegedly used a cannabis extract to alleviate cramps (Randall & O'Leary, 1998).

By the turn of the twentieth century, marijuana tinctures and extracts became more widely available. In the early 1900s, the Squibb Company offered a cannabis and morphine combination called Chlorodyne for stomach problems (Roffman, 1982). Labels from medical marijuana products from the beginning of the century claimed antispasmodic, sedating, analgesic, and hypnotic effects (Aldrich, 1997). By the 1930s, both Eli Lilly and Parke-Davis marketed such products (Mikuriya & Aldrich, 1988). Nevertheless, the Marijuana Tax Act of 1937, which required a special fee for the transport of marijuana, decreased medical use. The act required a prohibitive tax on the drug, which could reach $100 per ounce—quite a bit of money in 1937. (Details of this act appear in chapter 10's discussion of legal issues.) The *U.S. Pharmacopoeia*, which originally listed cannabis in 1850 as a cure for many ailments, removed the drug by 1941 (Bonnie & Whitebread, 1974). In the 1940s and 1950s, both medicinal use and research decreased markedly, particularly in the United States. One series of studies performed in Czechoslovakia in the late 1950s confirmed cannabis's antibiotic and analgesic effects (Kabilek, Krejci, & Santavy, 1960), but few other studies were published during this era.

For centuries, nearly every medicinal use of marijuana seemed comparable to the initial treatments described by Emperor Shen Neng. Finally, in the 1970s, data suggested a new medical application unlike any previously proposed—the treatment of glaucoma. This disorder, a leading cause of blindness, accompanies increased pressure inside the eye. Although successful treatments exist for the problem, some have odd side effects like blurred vision and headaches. In addition, a few individuals respond poorly to available medications and procedures. Some develop tolerance to the drugs, minimizing their ability to lower the pressure within the eye. A few patients receive surgery but show only small improvements. The discovery of marijuana's impact on intraocular pressure combined with these difficulties associated with other treatments suggested that cannabis may serve as a good medication for the problem.

The initial support for marijuana as a treatment for glaucoma appeared inadvertently in two different settings. A study at the University of California in Los Angeles (UCLA), designed to confirm police reports of cannabis-induced pupil dilation, found slight constriction of the pupil instead. Pupil constriction often accompanies decreased intraocular pressure, as it did in this case. Further work revealed the effect occurred for people with glaucoma as well (Grinspoon and Bakalar, 1997; Hepler & Petrus, 1976). At least one person with glaucoma made this discovery on his own—Robert Randall. The considerable force on Randall's optic nerve created the illusion of halos around lights. As innocuous as this symptom may sound, the halos signal the intense pressure that eventually leads to blindness. He had used all the treatments available at that time but found no relief. Once, after smoking marijuana, he noticed that the halos disappeared. His regular use of medical marijuana helped keep the pressure within his eye from creating total blindness. Nevertheless, he was arrested for marijuana possession and cultivation. He endured intense legal battles and submitted to considerable research. He finally became one of the few modern, legal users of marijuana for medicinal purposes (Randall & O'Leary, 1998).

Some of the latest developments in the history of medical cannabis and its derivatives concern the treatment of nausea and weight loss associated with chemotherapy and AIDS. Healers identified marijuana's antiemetic and appetite-enhancing effects at least by the 1500s (Abel, 1980). Many modern sufferers have turned to the drug for relief. Nabilone, a chemical derivative of one of marijuana's active components, was developed by Lily Research Laboratories and shows some of the same healing qualities as marijuana (Lemberger & Rowe, 1975). Unfortunately, this drug may prove toxic with continued use. It apparently builds up in brain tissue, preventing long-term prescription (Randall & O'Leary, 1998). Dronabinol, a synthetic version of THC sold under the brand name Marinol, mimics some of marijuana's therapeutic effects. This drug also increases appetite and decreases nausea.

Despite dronabinol's established positive effects, many argue for marijuana's superiority on medical and economic grounds. Patients prefer smoked marijuana to this medication. Anyone who is vomiting and nauseated may find swallowing a pill quite difficult. Because patients must digest the orally administered dronabinol, the effects do not appear as rapidly. Many claim that the dosage is much easier to modify with smoked marijuana, too. After a few puffs and a brief waiting period, patients can decide to increase their dose as they see fit. Dronabinol pills do not lend themselves to this sort of quick and easy alteration of dosage. The pills are also markedly more expensive. Patients could spend from $600 to over $1,000 per month on dronabinol; comparable doses of marijuana cost considerably less (Rosenthal & Kubby, 1996; Zimmer & Morgan, 1997).

The primary deterrent to using medical marijuana concerns legal sanctions. Marijuana remains a Schedule I drug, meaning it has no approved medical value. Doctors cannot prescribe it. Possession can lead to harsh penalties, including fines and imprisonment. In contrast, the FDA approved dronabinol for cancer patients in 1985 and for AIDS patients in 1992. In July 1999, the Drug Enforcement Agency even reclassified dronabinol as a Schedule III rather than a Schedule II drug, decreasing some of the paperwork and hassle required for prescribing it. Given marijuana's lower cost and potential efficacy, many have challenged its classification in Schedule I. Some physicians and organizations hope to help many suffering people and continue the history of medical marijuana by working toward reclassification (Grinspoon, 1971; Grinspoon & Bakalar, 1997).

A History of Cannabis the Intoxicant

The story of cannabis as a medicine is separate from its role as a recreational drug. The consumption of cannabis solely for psychoactive, mind-altering effects differs from other forms of use. Industrial hemp alters consciousness only in those who find fashion a source of ecstasy. Ailing patients who use medical marijuana report altered thoughts and feelings, but the alleviation of symptoms seems the primary goal. Some individuals use cannabis as part of religious ritual, with varying emphasis on intoxication. Recreational use of the drug relates to many artistic, religious, legal, and economic factors. The history of this form of cannabis consumption reveals a great deal about humankind's reactions to novelty, pleasure, and the unknown.

Given the origins of hemp in Asia, recreational cannabis consumption may have developed there first. Patients who ingested marijuana as medicine probably experienced its psychoactive effects. Differentiating medical and intoxicating use essentially depended on the presence of an illness. Emperor Shen Neng's patients back in 2737 B.C. may have consumed the plant even in the absence of symptoms. Ancient Chinese who ingested cannabis for fun rather than relief from sickness would qualify as the first recreational users. Cannabis was not the earliest known intoxicant. People had been drinking alcohol since at least 6400 B.C., and perhaps since 8000 B.C. (Mellaart, 1967; Roueche, 1963). The Assyrians and Sumerians had used opium for roughly a thousand years before Shen Neng (Scott, 1969), though it may not have reached China until 800 A.D. (Maisto et al., 1995). Industrial hemp preceded these drugs, but cannabis the intoxicant did not.

Views of the new intoxicant are not well documented for a couple of millennia after its first ingestion. Eventually, attitudes about the drug varied across different eras and locales, much as they do today. Around 600 B.C., China's Taoist movement grew more popular. Taoists of the day disapproved of cannabis (Abel, 1980). This initial condemnation preceded Asia's iron age and even the development of the metal plow (Garraty & Gay, 1981). Perhaps because the drug was not widely known outside of China, it was not condemned elsewhere. To other ancients, cannabis intoxication seemed more an oddity than a sin. Around 450 B.C., Herodotus, the Greek father of historical narrative, mentioned the drug in his riveting chronicle *The Histories*. He described a Scythian funeral rite that included inhaling the smoke from cannabis seeds (Herodotus, 1999). The Soviet archeologist Rudenko found evidence that confirms this practice (Rudenko, 1970). Scythians may have indulged in this intoxicant even between funerals, but Herodotus does not mention it. Much later, in Greece, Plutarch (47–127 A.D.) described Thracians throwing the tops of plants on the fire, huffing the fumes, and falling asleep. This plant was likely cannabis. Neither Herodotus nor Plutarch condemned the practices of these people; they simply reported on the habit of another culture.

China's attitudes about cannabis eventually changed. Around Plutarch's time, in 100 A.D., the Taoists began dabbling in alchemy. They purportedly used cannabis to induce hallucinations (Abel, 1980). This effect actually occurs only rarely and only at extremely high doses (Tart, 1971). The Taoists must have been quite motivated, though, because the cannabis visions supposedly revealed the path to immortality (Needham, 1974). The ancient physician Meng Shen wrote that visions required 100 consecutive days of consumption of the seeds (Li, 1974), suggesting that perhaps many users failed to hallucinate on average doses. Even this recommendation seems misleading; the seeds contain little THC. Perhaps they were covered with resin. Otherwise, eating pounds and pounds would not produce a psychoactive effect.

Although intoxication was clearly the goal, this Taoist use of marijuana may qualify as religious rather than recreational. Cannabis use in India developed a more formal connection to ritual and worship. The plant may have spread to India from China, but legends there suggest that the Hindu deity Shiva brought it down from the Himalayas (Abel, 1980). Devotees believe Shiva enjoyed cannabis; many offer it to the god on special festival days. By 2000 B.C., the substance appeared in religious texts. The holy book *Atharvaveda* mentions the plant with considerable praise, emphasizing its ability to reduce tension. Despite the myriad antianxiety medications available today, Westerners may label cannabis's anxiolytic effects as intoxicating rather than medicinal or religious. Nevertheless, consumption in India had and continues to have a religious quality.

Natives of India developed distinctions among cannabis products. People there use ganja, charas, and bhang. Ganja, the flowering tops, and charas, the plant's resin, are smoked in clay pipes. Natives eat bhang or use it to make a cold, liquid refreshment. The drink is also called bhang or thandai. Although recipes for the beverage vary, most contain cannabis, nuts, milk, sugar, poppy seeds, and at least a half dozen other spices. Bhang consumption has important religious associations; many people drink it at holy festivals. The sanctions against alcohol may have increased bhang's popularity in Hindu society. The drink is not particularly potent; digesting the cannabis may decrease some of its effects. Not all uses in India are purely sacramental. Hosts offer the beverage to guests much the same as Westerners might offer alcohol (Abel, 1980).

Hashish probably developed later. This concentration of the cannabis plant's resin apparently reached Arab countries by the year 1000. Otherwise, Ibn Wahshiyah, a physician of the era, would have had no need for his concerned but inaccurate warnings that the drug could be fatal. Despite evidence to the contrary, one legend of the invention of hashish remains particularly popular. This story suggests that Haydar, the Persian father of the Sufis, discovered hashish in 1155 (Rosenthal, 1971). Ibn Wahshiyah's earlier warnings belie the legend, but hashish is still called "the wine of Haydar."

Hashish's concentration of resin contains a higher percentage of psychoactive THC than other cannabis preparations. It maintains its potency longer than the plant, improving storage, shipment, and sales. The procedure for concentrating the resin has developed a few legends of its own. Different tales describe naked slaves or nubile women running through cannabis fields and scraping the resin that attaches to their bodies into cakes of hash. As romantic as that process may sound, shaking the plants and pressing the loosened resin together was probably simpler. Modern recipes describe combining cannabis oil with powdered marijuana, which is probably the contemporary technique of choice (Gold, 1989). This new product likely facilitated the spread from India to the Arab countries. Hashish traveled better than the whole plant and may have brought a higher price. This new location in the Arab world brought out one of the most long-lived and pernicious legends about the drug's link to violence.

Around the year 1090, Hasan-ibn-Sabah and his followers struck fear in the hearts of many. Hasan's devotees allegedly fulfilled his every wish, including murdering his enemies. Their loyalty allegedly stemmed from a belief that completing their missions guaranteed entry into paradise. Hasan may have offered them a glimpse of this nirvana in an effort to validate this claim of a blissful afterlife. One legend told that new initiates were drugged, blindfolded, and taken to a lush garden filled with various exotic amusements. After a few moments they were again drugged and blindfolded and removed from the garden with the promise that they would return at their deaths if they did Hasan's bidding. This experience apparently motivated the followers to act as instructed.

Over a hundred years later, Marco Polo returned from the Middle East with this tale. Odd alterations eventually crept into the story. The unnamed sedative became hashish. Later permutations suggested that the devotees used the concentrated resin immediately prior to murdering Hasan's enemies. These tales led many to believe hashish intoxication caused hostile acts. Some asserted that the name given these murderous followers of Hasan even derived from the name of the drug; the killers, who supposedly had consumed hashish, were called "assassins." Although "hashish" and "assassin" sound alike, other origins of the word may seem more tenable. "Assassin" may have originally meant "follower of Hasan." It may have developed from "hassass"—an Arabic word that means "to kill." Despite these compelling alternative etymologies, the connection between hashish and assassin stuck (Casto, 1970). The first head of the Federal Bureau of Narcotics, Harry Anslinger, even mentioned the story as evidence that cannabis incited crime (Bonnie & Whitebread, 1974). Data do not support the connection. Although the belief that cannabis causes violence remains (Schwartz, 1984), laboratory research reveals that marijuana actually decreases aggression (Myerscough & Taylor, 1985).

From Marco Polo's day through at least 1700, a compilation of stories from the Middle East known as *1,001 Nights* gained popularity. One anecdote in the book depicts hashish intoxication. A man ingests the drug at a public bath and fantasizes about a sexual encounter. Though he realizes he is only experiencing the drug-induced reverie, he finds himself tossed from the bathhouse for showing an obvious sign of arousal. Although the tale is hardly happy, it may have increased interest in the drug. The simultaneous experience of intoxication and awareness of that intoxication would later appeal to Europeans in search of new ways to enhance their creativity. The sexual nature of the story may have added to cannabis's reputation as an aphrodisiac, too.

Hashish's path from the Arab countries to European artists was convoluted. Napoleon's invasion of Egypt in 1798 exposed French troops to the drug. Napoleon outlawed all cannabis use, but soldiers and scientists returned to France with hashish. The book *1,001 Nights* had grown quite popular in all of Europe, and many knew its depiction of intoxication. Some artists at the time hoped that the cannabis experience might inspire new work. Perhaps the agony of writer's block, or the inability to paint, led artists to try any remedy. Few reports at the time mentioned addiction to hashish, so many experimenters decided to try the drug.

The French physician Jacques Joseph Moreau remains the most-cited connection between cannabis and the art community. Moreau first used hashish while traveling through the Middle East in the 1830s. He hypothesized that cannabis-induced sensations might model the hallucinations and delusions common in psychotic individuals. He had hoped this research might help treatment of the mentally ill (Moreau, 1845/1973). Timothy Leary, among others, later offered comparable conjecture about LSD and the hallucinogens (Leary, 1997). Moreau initially ingested hashish himself. He found the intoxication paralleled some aspects of psychosis. These results led Moreau to search for volunteers who might take the drug while he observed from a less intoxicated, more objective state. The outspoken hedonist and popular novelist Pierre Jules Theophile Gautier assisted Moreau in this research. He not only participated himself, but he also recruited other members of France's artistic community. This crew of experimenters donned the name "The Hashish Club" and met monthly in an old mansion in Paris.

Gautier published details of his first hashish experience in 1843 in his *Hashish Club*. The manuscript contained the drama and flair typical of French literature of the era. The night was stereotypically dark and cloudy. Everything inside the old mansion appeared "gigantic," "flamboyant," "dazzling," and "mysterious" (Gautier, 1846/1966). Gautier ingested dawamesc, a confection containing hashish, sugar, and spices. His hallucinations were markedly more elaborate than those reported by others who had used the drug. The writer may have exaggerated for effect. He also may have received a large and potent dosage of the drug (Solomon, 1966). Evidence suggests some hashish of the day contained opium, which may have altered Gautier's experience dramatically (Bell, 1857). Although much of the description sounds innocuous, other parts reveal paranoid, frightened, and sad reactions. Moreau's book based on this research appeared in 1846 and received an honorable mention in a scientific competition sponsored by the French Academy of Science (Abel, 1980).

The Hashish Club continued to meet, bringing Baudelaire, Balzac, Dumas, and Flaubert into its ranks. The impact of the drug on their creativity remains unknown. Modern studies have produced mixed results for marijuana's impact on originality (Chait & Pierri, 1992). A couple of members of the club wrote about the drug itself. Baudelaire published *On the Artificial Ideal,* a monograph about cannabis, in 1858. The book described the elaborate changes in thoughts and sensations that arise from the drug. He clearly mentioned both euphoric and dysphoric reactions, including some paranoid, terrifying moments typical of extreme dosages. He emphasized the importance of the setting in determining the drug's effects. This work also depicts synaesthesia, a confusion of one sense for another. For example, Baudelaire described the sound of color and the color of sound. This synaesthesia experience is more common with hallucinogen intoxication. He added a translation of De Quincey's *Confessions of an Opium Eater* and republished the work as *The Artificial Paradises* in 1860 (Solomon, 1966). His unfavorable opinion of these drugs appears in his subsequent book on hashish and opium entitled *The Flowers of Evil.*

Other famous members of the Hashish Club apparently devoted less time to the drug than Baudelaire or Gautier. They certainly published less on the topic. Balzac apparently only observed at the meetings. He sampled the drug rarely and wrote little about it, except for a letter that describes some mild effects (Balzac, 1900). Alexander Dumas also reportedly never used the drug excessively. He has no works devoted to cannabis, but his *Count of Monte Cristo* contains a passage remarkably reminiscent of the story of Hasan's assassins. Flaubert also supposedly only observed at the club. (One wonders who did ingest the drug while all of these famous folks watched.) None of his works focus directly on cannabis, either. At his death in 1880, he left notes for the novel *La Spirale,* which describes a man's degeneration from hashish use (Abel, 1980).

Cannabis spread across the English Channel to Britain in the 1800s, but it seemed to cause little commotion. In 1855, a member of Parliament confessed to using the drug during a trip outside the country (Urquhart, 1855). Later articles in English journals and magazines argued against some of the drug's purported negative consequences. The English writer Laird-Clowes was particularly outspoken against the idea that cannabis caused violence, suggesting that the effect was inconceivable (Laird-Clowes, 1877). English artists followed the French in consuming hashish for inspiration. Both William Butler Yeats and Oscar Wilde may have used the drug, though both preferred other intoxicants. However, neither focused large works on cannabis the way Baudelaire or Gautier had (Abel, 1980).

Eventually, cannabis the intoxicant reached the United States. The first mention of the drug by an American author appears in a poem by John Greenleaf Whittier in 1854 (Whittier, 1854/1904). Bayard Taylor, a popular American writer of the time, discusses his own use of hashish in two travel books published in the mid-1850s (Taylor, 1854, 1855). Although his initial experience with the drug in Egypt sounds innocuous, a subsequent ingestion of a large dose appears quite aversive. Taylor's tales inspired the young American Fitz Hugh Ludlow to experiment with the drug and anonymously publish *The Hasheesh Eater* in 1857 (Ludlow, 1857). After a couple of ingestions of low doses produced no effect, Ludlow ate quite a bit of the drug and had a negative reaction. Despite the adverse effects, Ludlow took the drug again within a couple of weeks. He described the synaesthesia that Baudelaire reported, as well as laughter, dry mouth, and some uncomfortable, anxious feelings. Ludlow used the drug repeatedly, reporting some genuine difficulty maintaining abstinence. Eventually, with a doctor's help, Ludlow abandoned all use. His book remains one of the most popular American texts on cannabis from this era (Abel, 1980). Ludlow's book may have inspired a few other Americans to experiment with the drug, but use was not high in the late 1800s. Opium, alcohol, and cocaine were more popular. All of these drugs would eventually become illegal in the years to come.

Laws prohibiting opium appeared by the end of the century, perhaps as a reflection of racist sentiment against Asians. Exaggerated reports of cocaine's effects, particularly in people of African or Caribbean descent, contributed to legislation against this drug. Comparable discrimination against Mexican and African immigrants may have contributed to later cannabis prohibition (Musto, 1999). Few people in the United States actually used marijuana at the turn of the twentieth century, but those who did were not members of mainstream, Protestant, Caucasian society. Initial attempts to restrict cannabis with a federal mandate in 1911 failed. Many local governments, particularly in areas with extensive immigration, passed their own antimarijuana laws. Newspaper accounts of the day suggested that few readers had any familiarity with the drug. The Eighteenth Amendment, outlawing liquor, went into effect in 1920. Although alcohol prohibition decreased drinking, it increased marijuana consumption. People drank more coffee after liquor was outlawed, too (Brecher, 1972). In the late 1800s in India, a situation opposite to America's alcohol prohibition occurred. British taxes levied in India made forms of marijuana so expensive that many people turned to alcohol consumption (Abel, 1980).

After prohibition was repealed in 1933, marijuana became the target of government control. Sensationalistic stories linked violent acts to cannabis consumption. The reports often ignored tenable alternative explanations of the aggression, like alcohol consumption or mental illness. Many of the most outlandish stories appeared in newspapers published by William Randolph Hearst. Hearst purportedly had financial interests in the lumber and paper industries. He may have sought to eliminate competition from hemp (Herer, 1999). Harry Anslinger took these newspaper tales to Congress to argue for the 1937 Marijuana Tax Act. The act did not make the drug illegal, but it did put high taxes on it. Anslinger proposed harsh sentences for drug violations and worked hard to have them increased with the 1951 Boggs Act. Research had already undermined Anslinger's contentions that cannabis caused violence, so the Boggs Act included extreme penalties for possession of the drug because people at the time believed it led to heroin addiction. (Marijuana as a gateway to harder drugs is discussed in chapter 3.)

Marijuana use was still quite limited in the 1950s, but as the 1960s progressed, cannabis's popularity increased dramatically. The substance was no longer limited to minority users; college students of all ethnicities throughout the country reported using the drug. By the end of the 1960s, a commission appointed by President Lyndon Johnson found little evidence for the drug leading to heroin addiction or violence. Many citizens hoped for a repeal of marijuana prohibition, with the formation of the National Organization for the Reform of Marijuana Laws (NORML) in 1970. President Richard Nixon declined to take any steps toward legalization, despite recommendations from a fact-finding commission. Some local areas minimized penalties for possession in the 1970s, but many of these laws were repealed as drug attitudes altered during the administrations of Ronald Reagan.

Depictions of cannabis intoxication in the arts remained. Allen Ginsberg, the Beat poet, wrote an essay in favor of legalization during an experience of intoxication (Ginsberg, 1966). Although the grammar remains unconventional, the essay shows considerable fervor for the topic. Comedians, comic strips, and entertainers make reference to the drug. Television shows frequently allude to cannabis by depicting teens lighting incense and speaking in jocular non sequiturs. Although the drug has been illegal for over 60 years, almost 5,000 years of use appears to continue. In 1997 and again in 1999, approximately one-third of all Americans admitted to use of cannabis at least once in their lifetime (Department of Health and Human Services [DHHS], 1998; Substance Abuse and Mental Health Services Administration [SAMHSA], 2000). The number of future users will undoubtedly vary with the artistic, religious, legal, and economic factors that contributed to consumption in the past. Unfortunately, many people have no knowledge of the causes, consequences, and correlates of use. The remaining chapters of this book should help provide this information.

Conclusions

Industrial, medical, and recreational uses of marijuana increased its fame throughout the world. The many uses of the cannabis plant have a long history, beginning in 8000 B.C. when Taiwanese artists used fibers from the stem to decorate clay pots. Ancients eventually turned the fibers to rope and later weaved them into hemp fabric. By 100 B.C., the Chinese had used cannabis to make paper. These products spread across the ancient world. By 850 A.D., the Vikings had dragged the ropes with them to Iceland. In 1000, hemp ropes helped the Italian navy dominate the seas. The hemp crop was so important that British farmers were commanded to grow cannabis or pay fines. Kings ordered the American colonies to export the crop, but they used it to make rope and fabric of their own. People also used the seeds and their oil in various foods, developing nutritious recipes that remain popular today. Cotton and synthetic fibers have replaced some of these ropes and fabrics, but a new movement supports industrial hemp as a more ecological alternative to these products. Contemporary merchants still sell shirts, shoes, and even hammocks made of hemp. The oil of the seed also appears in modern shampoos, soaps, and salves.

Medicinal marijuana first appeared in 2737 B.C. when the Chinese Emperor Shen Neng prescribed it for many ailments. These treatments grew more popular in all of Asia and down the coast of Africa. Religious uses developed in certain sects of Hinduism in India. By the 1500s, some Europeans had mentioned the plant's medicinal use. In the 1842, Irish physician William O'Shaughnessy published medical experiments that he conducted in India. Tinctures of cannabis appeared in pharmacies throughout the world. Physicians prescribed the drug for everything from earache to nausea. Legislation against cannabis forced the medical community to withdraw it as a treatment by the 1940s. Yet the movement for medical marijuana continues. Studies performed in the 1970s revealed a new potential application of the drug in the treatment of glaucoma. Research on smoked marijuana, THC, and other cannabinoids, though often hindered by bureaucratic difficulties, continues to reveal potential pharmaceutical applications.

The recreational use of cannabis likely followed its early prescription for physical ailments. The intoxicating effects of the plant may have contributed to the ancient Scythian practice of huffing the smoke during funerals, as first reported in 450 B.C. By 100 A.D., the Chinese Taoists used the drug to induce visions. The intoxicant inspired stories of sexual arousal, like the one in *1,001 Nights*, which appeared by 1200 A.D. News of the drug's psychoactive properties spread throughout Europe. Napoleon's soldiers brought hashish to France from Egypt in 1798. Moreau, the French physician, supplied the drug to many French artists and writers in the middle of the 1840s. Literary work about hashish contributed to experiments in the United States, where Fitz Hugh Ludlow published a tale of his intoxication in 1857. Use did not spread in America until after the turn of the century. By the 1930s, the drug was illegal in every state. Despite this legislation, use increased in the 1960s and 1970s. Large organizations designed to alter legislation formed. Recreational use continues, with approximately one-third of Americans trying the drug at some time. The future of this controversial fiber, medicine, and intoxicant will depend on a complex interaction of its biological, psychological, and societal effects.

Appendix: Timeline for Highlights in the History of Cannabis

8000 B.C.: Hemp fiber first appears in Taiwan.

2737 B.C.: Emperor Shen Neng of China first prescribes medicinal marijuana.

2000 B.C.–1400 B.C.: Indian holy book *Atharvaveda* mentions marijuana's antianxiety effects.

600 B.C.: Hemp rope appears in southern Russia.

450 B.C.: Herodotus's *Histories* mention hemp fabrics and Scythian use of cannabis as an intoxicant.

200 B.C.: Hemp rope appears in Greece. Chinese *Book of Rites* mentions hemp fabric.

100 B.C.: First evidence of hemp paper, invented in China (see 105 A.D.).
23–79 A.D.: Pliny the Elder's *The Natural History* mentions hemp rope and marijuana's analgesic effects.
47–127: Plutarch mentions Thracians using cannabis as an intoxicant.
70: Dioscorides, a physician in Nero's army, lists medical marijuana in his pharmacopoeia.
100: Imported hemp rope first appears in England.
105: Legends suggest Ts'ai Lun invents hemp paper in China at this time (see 100 B.C.).
130–200: Greek physician Galen prescribes marijuana medicinally.
200: First pharmacopoeia of the East lists medical marijuana. Chinese surgeon Hua T'o uses marijuana as an anesthetic.
300: Young woman in Jerusalem receives medical marijuana during childbirth.
570: French queen Arnegunde is buried with hemp cloth.
850: Vikings take hemp rope and seeds to Iceland.
900: Arabs learn techniques for making hemp paper.
1000: Hemp ropes appear on Italian ships. Arabic physician Ibn Wahshiyah's *On Poisons* warns of marijuana's potential dangers.
1090: Hasan-ibn-Sabah recruits assassins with hashish.
1155: Haydar allegedly invents hashish.
1200: *1,001 Nights*, an Arabian collection of tales, describes hashish's intoxicating properties.
1300: Ethiopian pipes containing marijuana suggest drug has spread from Egypt to the rest of Africa.
1532: French physician Rabelais's *Gargantua and Pantagruel* mentions marijuana's medicinal effects.
1533: King Henry VIII fines farmers who do not raise hemp.
1563: Portuguese physician Garcia da Orta reports marijuana's medicinal effects.
1578: China's Li Shih-Chen writes of antibiotic and antiemetic effects of marijuana.
1600: England imports hemp from Russia.
1616: Jamestown colonists grow hemp.
1621: Burton's *Anatomy of Melancholy* suggests marijuana may treat depression.
1753: Linnaeus classifies *Cannabis sativa*.

1764: Medical marijuana appears in *The New England Dispensatory.*

1776: Kentucky begins growing hemp.

1783: Lamarck classifies the plant *Cannabis Indica.*

1794: Medical marijuana appears in *The Edinburgh New Dispensary.*

1798: Napoleon's soldiers learn of cannabis and hashish in Egypt.

1842: Irish physician O'Shaughnessy publishes cannabis research in English medical journals.

1843: French author Gautier publishes *The Hashish Club.*

1846: French physician Moreau publishes *Hashish and Mental Illness.*

1850: Cannabis added to *The U.S. Pharmacopoeia.*

1854: Whittier writes first American work to mention cannabis as intoxicant.

1857: American writer Ludlow publishes *The Hasheesh Eater.*

1858: French poet Baudelaire publishes *On the Artificial Ideal.*

1890: Sir J. R. Reynolds, chief physician to Queen Victoria, prescribes medical marijuana.

1924: Russian botanists classify *Cannabis ruderalis.*

1937: Marijuana Tax Act passes, requiring special fees for prescriptions of the drug.

1941: Cannabis removed from *U.S. Pharmacopoeia.*

1951: Boggs Act increases drug penalties.

1960: Czech researchers confirm antibiotic and analgesic effects of cannabis.

1970: National Organization for the Reform of Marijuana Laws (NORML) forms.

1971: First evidence suggesting marijuana may help glaucoma appears.

1975: Nabilone, a cannabinoid-based medication, appears.

1985: FDA approves dronabinol, a synthetic THC, for cancer patients.

1992: FDA approves dronabinol for AIDS-wasting syndrome.

1999: Hawaii and North Dakota attempt to legalize hemp farming. DEA reclassifies dronabinol as a Schedule III drug, making the medication easier to prescribe.

2000: Legalization initiative in Alaska fails.

2001: Canada adopts federal laws in support of medical marijuana.

2

Cannabis Use and Misuse

This chapter describes the use and misuse of cannabis. Estimating the number of marijuana smokers proves difficult. Many people show an understandable reluctance to confess to illegal behavior. This reluctance may also bias estimates of the amount of cannabis consumed. Marijuana lacks the standard dosage common to many other drugs, such as a pack of cigarettes or a shot of whiskey. In addition, people can misremember, exaggerate, or minimize their use. Many different terms have developed to describe drug problems. Definitions of various kinds of misuse prove troublesome. The term "addiction" has no universal meaning. "Dependence" has a fairly specific meaning separate from abuse, but two people with marijuana dependence need not share a single symptom. "Abuse" has a formal definition that lacks precision. For example, two smokers who qualify for abuse may not share any of the same problems. Some people define any use of an illegal drug as abuse. Perhaps the best approach to defining misuse relies on cataloging individual problems that stem from the drug. This approach may provide the most specific information for treatment. Each of these issues appears in detail below.

Estimating Cannabis Use

Between 200 and 300 million people worldwide report smoking marijuana (Woody & MacFadden, 1995). Cannabis remains the most widely consumed illicit drug in Canada (Russell, Newman, & Bland, 1994), as well as the United States. In 1999, approximately one-third of U.S. adults (76 million) reported smoking marijuana at least once. About 9% of the U.S. population reported smoking marijuana in the previous year; 5% had used in the previous month (SAMHSA, 2000). This 1999 survey did not assess weekly users, but in 1996, approximately 3% of Americans had used the drug on more than 51 days in the previous year (DHHS, 1998). More men report using the drug than women. Approximately 6.5% of the females and 10.5% of the males age 12 and over reported smoking marijuana in the previous year (Greenfield & O'Leary, 1999).

The popularity of the drug also varies with age. An alarming 19% of Americans age 12–17 had already tried marijuana. Young adults age 18–25 often have the highest percentage of users. In 1999, 52% of the people in this age group had tried the drug at least once, and 17% had smoked in the last month. Fewer people above age 25 smoke cannabis, with rates approaching 40% for lifetime use (SAMHSA, 2000). Rates of use also change in different eras. From 1972 until 1979, the percentage of users in the United States increased steadily each year. At the peak in 1979, 68% of the people age 18–25 reported using the drug at least once. Rates have decreased since then. Only 51% of young adults had tried the drug in 1991 and only 44% in 1996 (NIDA, 1991; SAMHSA, 1997). Currently, most people do not use the drug or do not use it often. In the United States in 1999, nonusers outnumbered users in all age groups. The percentage of people who use weekly (less than 3%) or monthly (5%) appears relatively small.

These estimates require cautious interpretation. Although researchers gather data carefully, these surveys invariably rely on self-reports. These numbers may not represent the true number of marijuana smokers. Instead, they depict the percentage of people who are willing to admit to using the drug. Although even presidential candidates will acknowledge smoking marijuana, people often lie about illegal or socially undesirable behaviors (LaBrie & Earleywine, 2000; Wimbush & Dalton, 1997). The bias in reporting may stem from the legal and social sanctions associated with drug use. The amount of bias may differ depending on the age of the respondent and the era of the response. For many reasons, teenagers in the 1970s might have claimed to have used cannabis even when they had not. Peer pressure and general attitudes about marijuana at the time may have contributed to overreporting.

For other reasons, parents in the 1990s might have claimed they never used the drug when they actually had. They may worry that their children will use the drug, which may contribute to underreporting their own use. These sorts of biases make estimates inaccurate. Everyday forgetfulness can decrease the validity of reports, too. For example, infrequent users may not recall correctly if they used the drug in the previous month. Thus, these estimates can only provide a general feel for the number of marijuana smokers over the years. Even urine screens or hair tests could not reveal exactly how many people ever used the drug because these procedures cannot assess consumption in the distant past.

Although assessments of the number of users are difficult, estimates of the amount used prove even more cumbersome. Legal drugs like alcohol, caffeine, and nicotine have standard units for consumption. A shot of whiskey, cup of coffee, and pack of cigarettes may not vary dramatically. In contrast, no standards exist for amounts of marijuana. Pipe bowls and joints range considerably in size. Different plants also show a wide range of potency. The subjective experience achieved with any amount may vary dramatically, too (Zimmer & Morgan, 1997).

One approach to measuring the amount of consumption relies on estimates of the total amount of marijuana smoked annually. The amount of cannabis consumed per year in the United States is obviously difficult to guess. The Drug Enforcement Agency reports seizing 2,035 metric tons of marijuana, which they estimated as 10 to 15% of the total traffic. A metric ton is roughly 2,200 pounds, so marijuana consumption in a year would be roughly 4,477,000 pounds. If 9% of Americans use cannabis each year, they would smoke roughly 3 ounces per year per person. Obviously, the variation among smokers would remain high, with some smoking markedly more and some smoking markedly less than 3 ounces in a year.

Another approach to estimating amounts of cannabis consumption works backward from the number of users. Assume that each user smokes 1 gram of marijuana per occasion. Focusing only on daily users, these numbers suggest Americans consume approximately 1,476 metric tons of marijuana, roughly 3.25 million pounds per year. This weight translates to only 2 ounces per user per year (How Much Marijuana?, 1995). These rough averages are unlikely to profile any individual user; some people use more often and in greater amounts than others. Nevertheless, this estimate of 2 to 3 ounces per year per user provides a rough approximation to the truth.

These guesses at the number of users and the amounts they consume provide some sense of the popularity of the drug. Estimating the extent of marijuana problems, however, requires additional information. Several different approaches to the misuse of drugs have developed. Distinctions among these forms of misuse have created heated debates and misunderstandings. Generally, the definitions all focus on negative consequences rather than the amount consumed or the frequency of smoking. The choice of definitions for misuse has important implications. One view of misuse may make marijuana problems appear more common than another. In addition, the different conceptualizations of marijuana problems may lead to different approaches to interventions.

For example, stating that a drug creates a horrid addiction might suggest that the best treatment would require abstinence. This approach is typical of attitudes about heroin. In contrast, thinking of a drug as creating a mild nuisance might imply that decreasing use could alleviate the problems, but complete abstinence may not be necessary. People experiencing problems with caffeine often adopt this approach. Each idea of misuse has its own associated set of debates. Popular definitions of misuse include addiction, dependence, abuse, and problems. Each of these appears in detail below.

Definitions of Misuse

Addiction

Marijuana addiction proves difficult to define. Some researchers claim that marijuana is not particularly addictive. Experts assert that cannabis's addictive power parallels caffeine's (Franklin, 1990; Hilts, 1994). Hilts asked two prominent drug researchers to rank features of six common drugs: nicotine, caffeine, heroin, cocaine, alcohol, and marijuana. Both experts ranked marijuana last in its ability to produce withdrawal, tolerance, and dependence. Another study had experts rank 18 drugs on how easily they "hook" people and how difficult they are to quit. Marijuana ranked 14th, behind the legal drugs nicotine (ranked first), alcohol (ranked 8th), and caffeine (ranked 12th). Only hallucinogens (MDMA, mushrooms, LSD, and mescaline) ranked lower than marijuana (Franklin, 1990).

These results only reflect expert opinions, but other evidence suggests that marijuana is not particularly addictive. For example, only a fraction of those who try marijuana eventually use it regularly. Nevertheless, some users still develop troubles related to the drug, and many request assistance in limiting their consumption (Roffman et al., 1993). In the face of these problems, the low ratings of addictive propensity seem confusing and may arise from diverse meanings for the word addiction.

The term "addiction" developed to describe the repetition of a habit. Addiction initially did not necessarily involve drugs. Its Latin root, "addictus," means "state, proclaim, or bind." The origin suggests an obvious, stated connection between addicted people and their actions. The word connotes surrender and implies that an activity or substance has bound the person (Lenson, 1995). Addiction was usually treated as a bad habit, similar to picking one's nose compulsively. At the beginning of the twentieth century, at least in the United States, the term changed from a description of actions to a medical condition. This distinction may seem subtle, but converting a bad habit into a physiological disorder brings it into the domain of medical intervention. This medical approach implies that addiction is not just a troublesome activity; it is a personal condition. Physicians have transformed many troubles into biological illnesses, with many repercussions. This tendency to reframe personal difficulties as physiological deficits may give medical communities more power and money (Foucault, 1973; Szasz, 1961). The implication that drug problems require a biological intervention may increase the sales of medicines or hospital treatments.

Some medical texts support the term "addiction" as the proper expression for drug problems. This definition emphasizes preoccupation with the substance, compulsive use, and frequent relapses. People who spend considerable time and effort trying to obtain the drug appear preoccupied. Those who drive long distances or wait many hours to meet a connection might qualify. Those who frequently reminisce about previous intoxication or fantasize about future consumption also might qualify as preoccupied. Circles of friends who can discuss nothing but their current, previous, or future drug use obviously miss out on quality relationships because of a preoccupation with the drug.

Compulsive use describes the subjective sense that one is forced to consume the drug. It need not mean intoxication at every moment. Compulsive use also can include consistent consumption under identical circumstances, such as smoking marijuana at the same time each evening. People who cannot watch a movie or have sex without cannabis might qualify as compulsive users. Repeated use despite attempts to stop also typifies this definition of addiction. Proponents of this approach to defining problems emphasize loss of control. Loss of control implies that the initial use of the substance impairs the ability to stop. A tacit assumption in some medical settings suggests that these symptoms arise from a biological process, an interaction of a foreign chemical with internal physiology (Miller, Gold, & Smith, 1997). This approach may have inspired the disease model of addiction.

THE DISEASE MODEL OF ADDICTION The disease model generates considerable emotion in many who investigate, treat, or experience drug problems. The controversy surrounding the model reflects the history of human reactions to personal difficulties. That history reveals that the disease model shows genuine progress over previous ideas about drug problems. Ancients often attributed personal troubles to evil spirits, as some people do today. Their treatments echoed this idea. For example, some who believed demons caused problems performed exorcisms. Some fans of the evil spirits model used a procedure called trephining. They drilled holes in the heads of those who suffered in order to release the dastardly devils within. The obvious drawbacks of this approach inspired the moral model of addiction.

A moral model of problems succeeded the idea of evil spirits in some areas. The moral model attributed troubles to ignoble thoughts, actions, or character. Adherents of this model provided treatments that included religious education and church attendance. This approach may not have appealed to all problem drug users, but it certainly beat trephining. Some adherents to the moral model suggested that those with drug problems were weak-willed, deficient, or sinful. The moral approach moved the initial source of the disorder. The evil spirits model posited that the problems arose from outside the individual. In contrast, the moral model implied that the source of the problems was inside the individual.

The disease model provided advantages over the moral model by asserting that drug problems served as symptoms of an illness. This illness led people, through no fault of their own, to the problematic consumption of substances. The disease model helped minimize blaming addicts for symptoms beyond their control. Few people fault others for contracting botulism or influenza. No one tells a diabetic to "snap out of it," or "use willpower" to combat symptoms. Yet individuals with drug problems may hear these expressions repeatedly. The disease model suggests that condemnation and commands waste effort that could be better spent on respectful therapy. This model underlies one of the most popular approaches to substance abuse treatment, the twelve-step program, which appears in more detail in chapter 11. Despite the success of these programs, many researchers prefer an alternative model.

Critics of the disease model certainly support respectful treatment. Nevertheless, they also suggest that viewing drug problems as a disease can have drawbacks. In an effort to minimize blaming people for addictive behavior, proponents of the disease model may have created another set of problems. The definition of disease has grown slippery. Addiction may not qualify because it does not parallel other illnesses. No bacteria or viruses lead to substance abuse the way they create anthrax or AIDS. Genes do not cause addiction in the direct way they produce Down's syndrome or hemophilia. The symptoms of cancer do not flare up in certain environments the way that craving for liquor may increase in a bar. Despite these facts, some advocates of the disease model treat addiction as a purely biological phenomenon. This emphasis on biology can exclude important economic, societal, and psychological contributors (Peele, 1998).

The opinion that drug problems reflect a medical disorder has certain drawbacks. The idea ignores social aspects of addiction, creates a dependence on medical treatments, and may lead to higher rates of relapse. Viewing addiction as a purely biological phenomenon minimizes established links between social class and drug problems (e.g., Armor, Polich, & Stambul, 1978; Miller & Miller, 1997). This approach may blind people to the potential for limiting drug problems through social change. A purely biological approach may also lead people to rely inappropriately on medications rather than psychological treatment. Changing personal behavior is often difficult. Changing society can prove even tougher. Prescribing medication for a disease is often easier. The disease model also may contribute to higher rates of relapse because of a central idea about loss of control. A belief in this symptom, which describes an inability to use a drug in small amounts without starting a binge, may actually increase relapse rates (Marlatt, Demming, & Reid, 1973; Peele, 1998).

Increases in the risk of relapse may serve as a prime example of a drawback associated with the disease model. Problem users frequently report that initial consumption of a drug invariably leads to using markedly more than they ever intended. Many assumed that a chemical process associated with the experience of intoxication impaired their ability to stop consumption. This loss of control became synonymous with addictive disease. Yet alcoholics surreptitiously given alcohol do not show signs of uncontrolled drinking. In contrast, alcoholics who believe they have consumed alcohol after drinking a placebo do show less control over their drinking (Marlatt et al., 1973). These results suggest that what people think is more important than what they consume. The belief that one has consumed alcohol can lead to a bigger binge than actually drinking alcohol without knowing it.

No one has performed a study like this one with cannabis users. Nevertheless, the way abstinent people think about using a little marijuana determines if they will go on to use a lot. In one relevant study, marijuana users in treatment reported about their relapses. Some used on a single occasion, considered it a "slip," and returned to abstinence quickly. Others considered the single use a sign of weak will or disease and ended up consuming markedly more (Stephens, Curtin, Simpson, & Roffman, 1994). These data suggest that this sort of loss of control likely arises from a psychological rather than a biological process. Many researchers view these data as evidence against the disease model.

Other definitions of both addiction and disease have added to the controversy. Peele (1998) emphasizes tolerance, withdrawal, and craving as essential to addiction. His work returns to the old definition of addiction, which can include actions that do not require chemicals. He extends the concept beyond drugs to nearly every behavior imaginable, including love (Peele, 1975). Yet he remains one of the most outspoken critics of the disease model. Tolerance, withdrawal, and craving all vary with features of the environment, suggesting that more than biology contributes to addictive behavior. Peele (1998) asserts that this evidence helps discredit the disease model. Other researchers argue that Peele misunderstands addiction (Wallace, 1990). The word may have so many different uses that it has lost its meaning. Thus, other terms have developed to describe trouble with drugs.

ALTERNATIVES TO THE CONCEPT OF ADDICTION Because many define addiction quite broadly and disparately, some mental health professionals prefer the terms "dependence" and "abuse." Others see these words as pejorative and judgmental compared to "addiction" (Miller, Gold, & Smith, 1997). Oddly enough, the World Health Organization (WHO) proposed the word "dependence" to avoid the derogatory aspects of the word "addiction" (Eddy, Halbach, Isbell, & Seevers, 1965). Addiction may imply a purely physical, biological process that might neglect psychological contributors to drug problems (Goldberg, 1997). Other terms have developed to focus on the observable behavior without hypothesizing an internal process or disease.

Focusing on observable behavior has been a recurring theme for the *Diagnostic and Statistical Manual* (*DSM*) developed by the American Psychiatric Association. This book attempts to define all psychiatric illnesses. Dependence and abuse appear in this work; addiction does not. Their definitions have gone through many revisions and probably will continue to do so. The first version of the manual (the *DSM-I*) appeared in 1952 (American Psychiatric Association [APA], 1952); it is now in its fourth edition. Originally, the opinions of many mental health professionals contributed to the definition of any disorder. Gradually, researchers attempted to clarify the diagnoses based on science rather than opinion. Early versions of the dependence diagnosis simply required "evidence of habitual use or a clear sense of need for the drug" (APA, 1968). This definition proved too subjective to diagnose reliably. Current definitions focus on a maladaptive pattern of use that leads to impairment or distress. Other symptoms are required for the diagnoses, as described below.

Dependence

The *DSM-IV* defines drug dependence as a collection of any three of seven symptoms. All must create meaningful distress and occur within the same year. The diagnosis requires a certain amount of judgment on the clinician's part, but the symptoms tend to be obvious. Each symptom reflects the idea that a person requires the drug to function and makes maladaptive sacrifices to use it. The current diagnosis focuses on consequences, not the amount or frequency of consumption. In contrast, earlier versions of the *DSM* once employed the frequency of intoxication as a symptom. For example, the diagnosis of a disorder known as "habitual excessive drinking" required intoxication 12 times per year (APA, 1968). This approach proved inexact and failed to relate to the magnitude of difficulties. Perhaps many problem drinkers purposely got drunk only 11 times a year to avoid this label. Thus, current diagnoses of drug dependence focus on negative consequences. They include tolerance and withdrawal, which were once considered the hallmarks of dependence. The additional symptoms are use that exceeds initial intention, persistent desire for the drug or failed attempts to decrease consumption, loss of time related to use, reduced activities because of consumption, and continued use despite problems.

Tolerance is one of the hallmarks of physiological dependence. It occurs when repeated use of the same dose no longer produces as dramatic an effect. This symptom can indicate extensive use and may motivate continued consumption. People do not grow tolerant to a drug but to its effects. After repeated use, some of the effects of a drug may decrease while others may not. Tolerance to the desired effects of marijuana may encourage people to smoke more. Many people report smoking cannabis to enhance their moods (Simons, Correia, Carey, & Borsari, 1998). Yet tolerance develops to the mood-enhancing effect of THC (Haney, Ward, Comer, Foltin, & Fischman, 1999a). This tolerance may lead people to smoke more to achieve the same emotional reactions. The increased use may coincide with a greater chance for problems. Ironically, tolerance to negative effects may also encourage more consumption. For example, smoking marijuana creates dry mouth, but this effect diminishes with use (Weller & Halikas 1982). This negative effect may have inhibited use initially. People might stop smoking if their mouths became too dry. But once tolerance develops, their mouths do not grow as dry and they may smoke more. Thus, tolerance to marijuana's effects may lead to increased consumption and serves as a symptom of dependence.

The second symptom of dependence is withdrawal. Withdrawal refers to discomfort associated with the absence of the drug. Many drugs produce withdrawal, including the most common ones: caffeine, nicotine, and alcohol. The most notorious drug withdrawal may come from heroin. This opiate has a reputation for producing dramatic withdrawal symptoms, including notorious leg twitches that may have inspired the expression "kicking junk." No two people experience withdrawal in the same way. Many assert that marijuana does not produce any withdrawal at all. It certainly does not create the dramatic symptoms characteristic of alcohol or heroin, and many users do not experience any problems after discontinuing use (Schuckit et al., 1999). Nevertheless, people given synthetic THC for a few consecutive days report negative moods and disturbed sleep after they stop taking the drug (Haney et al., 1999a). People who smoke marijuana a few days in a row report more anxiety without the drug (Haney, Ward, Comer, Foltin, & Fischman, 1999b). Thus, marijuana can lead to withdrawal and therefore dependence.

The lack of flagrant, obvious cannabis withdrawal symptoms inspired the American Psychiatric Association to distinguish between types of dependence. Early versions of the diagnosis of dependence specifically noted that marijuana might cause problems in individuals who do not experience withdrawal (APA, 1968). The *DSM-IV* distinguishes between dependence with and without a physiological component. If tolerance or withdrawal appear among the three required symptoms, a diagnosis of physiological dependence is appropriate. Nevertheless, even without tolerance or withdrawal, individuals may receive a diagnosis of substance dependence without a physiological component. If they show three other symptoms, they will still receive the diagnosis. This change in procedure has made the diagnosis of marijuana dependence potentially more common.

A third symptom of dependence involves use that exceeds initial intention. This symptom suggests that individuals may plan to consume a certain amount of a drug, but once intoxication begins, they use markedly more. Use that exceeds intention was once known as loss of control. Many people misinterpreted the idea of loss of control, suggesting it meant an unstoppable compulsion to use all of the drug available. People who got high and still had marijuana in the house the next morning might have claimed that they did not show loss of control. Use that exceeds intention specifically does not imply this dramatic, unconscious consumption. This symptom simply suggests that dependent users may have trouble smoking a small amount if they intend to. Ironically, people who never intend to smoke a small amount may not get the opportunity to qualify for this symptom.

The fourth symptom of dependence is failed attempts to decrease use or a constant desire for the drug. An inability to reduce marijuana consumption despite a wish to do so certainly suggests that the drug has altered behavior meaningfully. Yet someone with no motivation to quit would likely never qualify for a failed attempt. Thus, people who have not attempted to quit may still qualify for this symptom if they show a persistent, continuous craving for the drug. An inability to stop or a constant desire suggests dependence.

A fifth symptom of dependence involves loss of time related to use. The time lost can be devoted to experiencing intoxication, recovering from it, or seeking drugs. Because marijuana is illegal, users may spend considerable time in search of it. People addicted to caffeine, nicotine, or alcohol may prove less likely to lose time in search of their drug of choice. The number of hours required to qualify for a meaningful loss of time is unclear, making this symptom seem subjective. Clear-cut cases include anyone whose day is devoted to finding drugs, getting high, and recovering. Anyone who spends a few hours each day on these activities would also qualify. In contrast, individuals who smoke marijuana an hour before bed each night might argue that they have lost little time and should not qualify for this symptom. The subjective assessment of a meaningful amount of time may contribute to problems with the diagnosis of dependence.

The sixth symptom of dependence is reduced activities because of drug use. This symptom focuses on work, relationships, and leisure. The presence of this symptom suggests that the drug has taken over so much of daily life that the user would qualify as dependent. Any impairment in job performance because of intoxication, hangover, or devoting work hours to obtaining drugs would qualify for the symptom. Anyone who misses work each Monday to recover from weekend binges might qualify for reduced activities. Sufficient functioning at work, however, does not ensure against dependence. Even with stellar job performance, impaired social functioning can also indicate problems. If a user's only friends are also users, and they only socialize while intoxicated or seeking drugs, the substance has obviously had a marked impact on friendships. Recreational functioning is also important to the diagnosis. A decrease in leisure activities suggests impaired recreation. A smoker who formerly enjoyed hiking, reading, and theatre but now spends all free time high in front of the television would qualify for the symptom. This approach to the diagnosis implies that cannabis smokers who are not experiencing a multifaceted life can improve the way they function by using less.

The last symptom of dependence requires continued use despite problems. People who persist in using the drug despite obvious negative consequences would qualify for this symptom. Recurrent use regardless of continued occupational, social, interpersonal, psychological, or health trouble obviously shows dependence. Many of these difficulties involve meaningful others in the user's life. Continued consumption in the face of conflicts with loved ones, employers, and family might qualify for this symptom. This creates an odd diagnostic situation because the symptom may vary with the person's environment. These interpersonal conflicts may arise from different attitudes about drugs among the relevant people. For example, in a family that considers any illegal drug consumption problematic, an occasional user might experience many fights. Someone using equally often in a family with relaxed attitudes about drugs would experience less conflict. This situation supports the idea that anyone who continues to use despite negative consequences must have a strong commitment to the drug, but members of a drug-oriented subculture might be less likely to be diagnosed with this symptom. Other problems need not involve people in the smoker's life. For example, anyone with emphysema who continues smoking would qualify for this symptom. People who report guilt or a loss of self-respect because of their drug use also qualify for this symptom. Those who continue using, even when it leads them to have a negative view of themselves, show a genuine sign of dependence.

Abuse

A subset of individuals may experience negative consequences from drugs that do not qualify for dependence but still lead to the diagnosis of substance abuse. This diagnosis requires significant impairment or distress directly related to the use of the drug. This dysfunction and strain are necessary to identify abuse. The diagnosis requires only one of the four symptoms that appear in the current criteria (APA, 1994). These symptoms include interference with major obligations, intoxication in unsafe settings, legal problems, and continued use in the face of troubles. Each of these signs requires some interpretation on a diagnoser's part, but trained individuals apply the category reliably. Most experienced diagnosticians can agree who meets criteria for substance abuse and who does not (Uestuen et al., 1997). This definition remains distinctly separate from dependence, which requires different symptoms and more of them. Although a diagnosis of abuse clearly serves as a sign of genuine troubles, many clinicians consider dependence more severe. Thus, those who qualify for dependence would not receive the less severe diagnosis of abuse.

The first symptom of abuse, interference with major obligations, requires impaired performance at work, home, or school. The idea that abuse requires interference with major obligations reflects concerns about optimal functioning. The impairment may arise because of intoxication, recovery from intoxication, or time devoted to searching for drugs. The definition is necessarily broad in order to apply to people with a variety of responsibilities. This symptom applies to employees who miss work because of hangovers, students who fail tests because they attend class high, or parents who neglect their children to go buy dope. One curious aspect of this symptom concerns the artful way some potential abusers arrange their lives to minimize the impact of their drug use on obligations. Anyone with few major obligations may become intoxicated more often or more severely without qualifying for the symptom. People with vacation time and no childcare duties could spend nearly the entire period intoxicated without interfering with major obligations. Under these circumstances, these people would not qualify for the symptom. Yet people who smoke markedly less but miss work or neglect children clearly show interference with their obligations and qualify for the abuse diagnosis.

The second symptom requires intoxication in an unsafe setting. The *DSM* specifically lists driving a car and operating machinery as hazardous situations where intoxication could create dangerous negative consequences. Many experienced drug users claim that their intoxicated driving differs little from their sober driving. These statements may reflect poorly on their driving abilities in general, but data suggest that some people slow down and leave more space between cars if they drive while high (see chapter 9). At least one study found improved driving with intoxication (Smiley, 1986). Nevertheless, given marijuana intoxication's established impairment of cognition, the National Organization for the Reform of Marijuana Laws (NORML, 1996a) clearly states that driving after smoking is unacceptable. Driving a car high, even for only a few blocks, qualifies as substance abuse.

The intoxicated performance of any task can lead to this diagnosis if impairment might create negative consequences. These actions need not be as elaborate as climbing mountains or handling firearms. Driving a forklift or using power tools might qualify. Note that no negative consequences actually need to occur; their increased likelihood can qualify for abuse. Thus, those who drive high but never receive tickets or have accidents would still qualify for abuse because they have increased their likelihood of negative consequences.

The third symptom included in the diagnosis of substance abuse concerns legal problems. This symptom may say as much about society's values as an individual's behavior (Brecher, 1972; Grilly, 1998). The definition of this symptom makes users of legal drugs less likely to get a diagnosis of abuse than users of illegal drugs. For example, possession of alcohol rarely leads to legal problems that would qualify for abuse; possession of marijuana could. In a sense, those who willingly ingest a drug despite these legal sanctions must have considerable investment in the substance. Any arrest that arises from drug-impaired behavior, such as public intoxication or driving under the influence, clearly qualifies as abuse. Other legal problems qualify even if they do not arise from intoxication. These include possession, sales, and intention to distribute.

Legal problems are arguably the very worst negative consequences associated with marijuana. As discussed in chapter 10, federal penalties for cultivation of 100 plants include life imprisonment and a $10,000,000 fine (Margolin, 1998). No other consequence of marijuana use is so severe.

The fourth symptom of drug abuse concerns consistent use despite problems. This symptom is identical to the last symptom of dependence. Note that recurrent use in the face of occupational, social, interpersonal, psychological, or health troubles qualifies as abuse.

CRITIQUES OF DEPENDENCE AND ABUSE The abuse and dependence diagnoses may have some advantages over the term addiction. Unlike addiction, both abuse and dependence have widespread agreement on their specific definitions. Diagnosticians can apply the categories reliably, too (Uestuen et al., 1997). Nevertheless, these diagnoses, and all those in the *DSM*, have many critics. The criticisms tend to focus on three problems with diagnoses, in general. First, a political atmosphere influences diagnoses in ways that make them appear unscientific. Second, people diagnosed with disorders suffer from associated stigma. Third, the definitions may lead clinicians and drug users to minimize important problems that do not qualify for a diagnosis.

First, political agendas influence the definitions of diagnoses, suggesting that they are not as scientific as would be ideal. Proponents of the *DSM* argue that abuse and dependence are genuine phenomena that exist in nature. Therefore, refining the definitions should improve our ability to identify the disorders. This approach suggests that abuse and dependence existed all along and were discovered through scientific inquiry. With increased research we should find the sets of symptoms that define abuse and dependence perfectly. Critics of the *DSM* argue that abuse and dependence, as well as other disorders, are not genuine phenomena but social constructions. They see refinements in the definitions as a result of politics of an era and the opinions of psychiatrists in power. Thus, many assert that these disorders are not discovered, they are invented.

Critics of the *DSM* point to ways that revisions have coincided with political agendas. For example, people attracted to others of the same sex were once classified with a disorder in the *DSM* (APA, 1968). After concerted and difficult work on the part of many activists, this diagnosis has been removed. Clearly, the inclusion and exclusion of same-sex sexual attraction reflect a political process, not scientific inquiry. Just as homophobia contributed to this diagnosis, critics suggest that sexism has contributed to other categories. Premenstrual dysphoric disorder, a clinical description of the common notion of premenstrual syndrome, appears in the *DSM-IV*. Obviously, the category can only apply to women. Even if premenstrual syndrome exists, its inclusion as a mental disorder could have extensive negative consequences. It has already served as a legal defense to excuse criminal behavior. The diagnosis also may hurt women seeking employment. The proposed disorder may reflect more about the American Psychiatric Association's attitudes about women than it does about the existence of a natural phenomenon (Caplan, 1995; Kutchins & Kirk, 1997).

Critics of diagnoses also focus on the negative repercussions that arise from these labels. Once mental health professionals decide that a set of symptoms form a disorder, the label has implications of its own. Lawyers have argued that substance dependence or abuse excuses crimes like the murder of children. This inadvertent result of making drug problems into mental illnesses can hurt people seeking employment as well (Peele, 1998). The stigma associated with labels like "drug addict" or "drug abuser" may decrease an individual's chances of finding a job and exacerbate other problems.

Another potential disadvantage of these diagnoses concerns their implied distinction from drug problems. A subset of marijuana users may experience mild problems that may improve with modest interventions. Clinicians who rely too heavily on diagnoses may miss opportunities to intervene if people do not qualify for abuse or dependence. Recent work on reducing the harm associated with drugs suggests that perhaps focusing on addiction, abuse, or dependence may prove less helpful than directing efforts toward minimizing individual problems (Marlatt, 1998). Thus, focusing on problems may help avoid the drawbacks of diagnoses, including their variation with political agendas, labeling, and inadvertent minimizing of troubles that do not qualify for a diagnosis.

Problems

Describing drug-related difficulties as addiction, abuse, or dependence creates certain misunderstandings. All three words may not only sound deprecating (Eddy et al., 1965; Miller, Gold, & Smith, 1997), but they also lack clarity. Addiction has no universally accepted definition. Abuse and dependence have formal definitions and provide a shorthand to communicate sets of symptoms quickly and easily. Nevertheless, the words do not reveal an individual's actual troubles. Anyone who qualifies for abuse may have one or more of the four symptoms required. The term "abuse" could mean any one of over a dozen combinations of different symptoms. Dependence requires three of any seven symptoms, providing over 30 potential combinations of symptoms. The terms also may encourage minimizing problems that do not qualify for diagnosis, perhaps interfering with treatment.

People experiencing negative consequences from drugs may prove unwilling to limit consumption if they do not qualify for addiction, abuse, or dependence. These limitations of the terms addiction, abuse, and dependence have inspired an approach that emphasizes problems rather than diagnoses or diseases. Thus, instead of worrying about whether a specific user qualifies for a disorder, time might be better spent identifying individual problems related to marijuana use. For example, a client may report frequent fatigue. A survey of this person's drug use may reveal that the fatigue often follows a night of smoking marijuana. Although this problem may not interfere enough to qualify for abuse, the client may benefit from smoking less, smoking earlier in the evening, or quitting. This emphasis on problems may allow clinicians to avoid pointless arguments about whether or not someone is an addict. Instead, clinician and client can focus on reducing the harm marijuana may cause.

THE PREVALENCE OF PROBLEMS The frequency of marijuana problems may prove difficult to estimate. Just as people are unwilling to confess to using the drug, they may also be reluctant to admit to related troubles. Thus, any studies of self-reported marijuana problems may underestimate their number. One of the most comprehensive studies of abuse and dependence began with interviews of over 42,000 people. This research focused on people who had used cannabis in the previous year and revealed that 23% qualified for a diagnosis of abuse and 6% qualified for a diagnosis of dependence. Abuse appeared more often among rural users. Dependence appeared more often among users who were depressed (Grant & Pickering, 1998).

Other studies have concentrated on negative consequences rather than diagnoses. Recent, large-scale investigations focused on problems related to social functioning, health troubles, or psychological symptoms. For example, those who argue with friends and family about their cannabis consumption qualified as having a social problem. Health problems included any medical or physical symptom that the respondents attribute to marijuana. Psychological problems included sad mood or a loss of interest in activities. In a large sample of Americans, 85% percent of people who had smoked marijuana in the previous year reported none of these problems. Fifteen percent reported one, 8% reported at least two, and 4% reported at least three negative consequences that they attributed to marijuana. Thus, more than four out of five people who had smoked marijuana in the previous year reported no problems related to the drug, but 15% might improve their lives by limiting their consumption (NIDA, 1991).

This information certainly helps provide estimates of marijuana problems, but the data raise many questions, too. At first glance, it appears that 15% of marijuana smokers experience problems with the drug. Unfortunately, we have no idea how many people who did not use marijuana experienced comparable social, medical, or psychological troubles. A meaningful control group that included people who never smoked marijuana would certainly help interpretations of this study. People who have never used any drugs argue with their families, lose interest in activities, or have health problems, too. Some of the users in this study may have experienced these symptoms, even if they had never smoked marijuana. Yet the tacit assumption, that the marijuana created the problems, remains untested. If marijuana smokers reported more of these sorts of troubles than nonsmokers, the idea that cannabis caused the problems would receive some limited support. The current approach, however, may overestimate marijuana's negative impact.

The limitations of this one study do not mean that marijuana does not cause problems. Other research supports the idea that a percentage of marijuana users experience troubles with the drug. Approximately 9% of one group of smokers followed for 5 years developed negative consequences (Weller & Halikas, 1980). These researchers defined problems in four aspects of life. These included negative effects of the drug, problems controlling use, and interpersonal difficulties. They also included unfavorable opinions about use. The negative effects of the drug included physical health problems, blackouts, or a subjective feeling of dependence. Problems controlling use consisted of 48-hour binges, use in the early morning, or an inability to limit consumption. Interpersonal difficulties involved fights with friends and loved ones. Adverse opinions included feeling that marijuana use had grown excessive, guilt-inducing, or objectionable. Unlike the 1991 NIDA study, which focused on problems that could have occurred to anyone, this study identified troubles that concentrate more on marijuana. The 9% of the sample labeled problem users experienced troubles in at least three of these domains. These studies both suggest that marijuana use is not harmless, and that some individuals experience negative consequences from the drug. Even those who may not qualify for addiction, abuse, or dependence might benefit from altering their marijuana consumption. A focus on problems may enhance the prevention of addiction, abuse, or dependence, however they are defined.

Conclusions

Marijuana is the most commonly consumed illicit drug, with 200 to 300 million users worldwide. Approximately one-third of Americans have tried the substance at least once. Despite its popularity, few people smoke marijuana regularly. Less than 5% of Americans report using the drug every week. Estimating the number of users is difficult because people lie or forget about their use. The amounts that people consume are also hard to estimate. A variety of definitions of misuse of the drug exist that include addiction, dependence, abuse, and problems. Addiction does not have a universal definition, making the term difficult to use scientifically. Abuse and dependence are diagnosed reliably and clearly can apply to problem marijuana users. Nevertheless, the abuse and dependence diagnoses may not provide the clear information one might learn from a simple list of marijuana problems. Both diagnoses may say more about the culture and values of a given clinician than the actual negative consequences that cannabis creates. Marijuana problems are not particularly common, but 6 to 23% of users report some difficulties with the drug. Techniques for decreasing these marijuana problems appear in chapter 11.

Stepping-Stones, Gateways, and the Prevention of Drug Problems

Researchers, theorists, politicians, and parents have all expressed concern about marijuana's potential to lead to the use of drugs with worse negative consequences. Proponents of these stepping-stone and gateway theories suggest that even if the adverse effects of marijuana are minimal, the drug can still cause trouble by ushering users toward the consumption of other illicit substances, including heroin and crack. According to this premise, marijuana should remain a primary concern because these other drugs create so many hardships. Comparable arguments appear against underage drinking and cigarette smoking. The gateway and stepping-stone theories have generated considerable research and debate for many years. The research remains difficult to evaluate without clear definitions of a stepping-stone and gateway. Interpreting this literature requires a good understanding of causality. Many popular reports confuse the causes of drug use with simple precursors. Confusion about the actual causes of drug consumption can impair any effort to prevent substance abuse and related problems. Thus, this chapter defines a stepping-stone and gateway, reviews the requirements for causality, examines the literature relating marijuana consumption to the use of harder drugs, and discusses the prevention of drug problems.

A stepping-stone provides a helpful foothold along a path, which serves as an odd metaphor for drug use. Stepping-stone theories often imply that marijuana produces a biological effect that somehow leads to the uncontrollable consumption of other drugs. This sort of theorizing began over 40 years ago (Nahas, 1990). Descriptions in popular culture create the impression that marijuana intoxication produces an insatiable urge for more and different drugs, something similar to the way eating salt makes people thirsty. Data do not support these ideas. Marijuana and hard drugs do share some biological effects. For example, THC, the opiates, and cocaine alter the dopamine system in comparable ways (Koob & Le Moal, 1997). The cannabinoids, however, have their own receptor that does not react directly to drugs like heroin and cocaine.

Additional evidence against a biological stepping-stone appears in animal research. If marijuana created physiological changes that increased the desire for other drugs, animals exposed to cannabis would likely ingest other intoxicants when given the opportunity. Yet rodents exposed to THC do not show a sudden willingness to press levers for other drugs. They do not even appear willing to give themselves more THC (Schenk & Partridge, 1999; Wiley, 1999). Thus, physiological mechanisms do not explain any link between marijuana and the use of other intoxicants (Institute of Medicine [IOM], 1999; Zimmer & Morgan, 1997).

Gateways and Causality

The lack of evidence for an obvious, biological stepping-stone inspired theorists to formulate ideas about marijuana as a gateway drug. A gateway usually serves as a passage to a region. Proponents of gateway theory show that people who use drugs like heroin and cocaine often used alcohol, tobacco, and marijuana first (Kandel, Yamaguchi, & Chen, 1992; Miller, 1994). Confusion about these theories has led to the idea that gateway drugs cause users to consume other substances. The prominent researchers of gateway theory never state that one drug causes the ingestion of another one. They simply report that cigarette smoking often precedes marijuana consumption, which usually precedes the use of other illicit drugs. Nevertheless, a few other authors misunderstand these data and create the impression that smoking marijuana leads inevitably to the use of other drugs (Nahas, 1990). Avoiding these misunderstandings requires a thorough understanding of causes.

The idea that one drug causes the use of another drug is difficult to prove. Proof of a cause creating an effect requires at least three clear criteria. These criteria were first proposed in the 1700s by the Scottish philosopher David Hume, one of the British empiricists. Hume emphasized that a cause creates an effect only under certain conditions. The conditions are association, temporal antecedence, and isolation. Association means that the cause and effect must occur together. Temporal antecedence means that the cause must occur first, before the effect. Isolation means that all alternative causes must be ruled out. That is, the effect could not have occurred because of some other potential cause (Hume, 1739).

The best evidence for causality comes from experiments where the hypothesized cause can be manipulated. The cause should lead to the effect; its absence should create no effect (or a different one). For example, an experiment might reveal that marijuana causes intoxication by comparing people who smoked cannabis to an equivalent group of other people who smoked a placebo. This sort of experiment could provide evidence for all three of Hume's criteria. A large group of people who are not under the influence of any drug might begin by rating their intoxication. Presumably, all would claim their intoxication level was 0. The experimenter would then choose half these people at random to smoke marijuana. The other half would smoke a credible placebo. This random selection of people would help ensure that the two groups are similar. A second assessment of intoxication would likely reveal higher ratings of intoxication for the people who smoked marijuana than those who smoked the placebo.

The data from this experiment would satisfy Hume's criteria. The marijuana group's higher intoxication ratings reveal that the drug and the effect are associated. Because the groups did not differ before smoking, only after, the marijuana apparently preceded the effect. Thus, the temporal antecedence is fulfilled. Finally, all other sources of intoxication are ruled out because the two groups only differ in smoking cannabis or placebo. This condition helps eliminate any alternative explanations or causes. These data support the conclusion that marijuana can be isolated as the cause of this intoxication.

A simple experiment works well for proving cannabis causes intoxication, but researchers cannot use a human experiment to test the gateway theory. Randomly exposing people to marijuana to see who goes on to snort coke, bang junk, or drop acid creates many practical and ethical problems. These problems make isolating marijuana as the cause of other drug use impossible in humans. An obvious alternative approach could employ animal participants. No animal experiments have found that exposure to THC increases the likelihood of using other drugs or of even working for more THC (Schenk & Partridge, 1999; Wiley, 1999). Thus, gateway theory's only support comes from correlational studies.

Correlational investigations often examine users of crack or heroin and ask which drugs they used previously. These studies may establish an association and temporal antecedence, but fail to isolate marijuana as a cause. At first glance, the idea that users of hard drugs used marijuana previously seems compelling evidence for marijuana's contribution to drug problems. Nevertheless, a couple of exaggerated examples may illustrate how treating these data as causal evidence is erroneous.

Errors in Causal Reasoning

Suppose data revealed that the crime rate in a city rises as the number of churches increases. This association might lead some cynic to hypothesize that churches cause crime. Data may suggest that the churches are built prior to the increases in crime, further supporting the theory. Both the association between cause and effect and the precedence of the cause appear. Nevertheless, these two facts alone do not establish that churches cause crime. An alternative explanation remains. As cities grow larger, both crime and the number of churches increase. The size of the population accounts for both of these increases. One need not cause the other. Another example concerns shoe size and vocabulary. Data reveal that people who know more words also have larger shoes. One might hypothesize that memory for words is stored in the feet. Obviously, age can account for this relationship. As children grow older their feet grow and they learn more words.

Although these examples appear absurd, data misinterpreted to support that marijuana causes crack addiction are comparable to those in support of churches causing crime and vocabulary increasing shoe size. In fact, the correlations for these absurd examples are probably larger than those linking marijuana to crack. Most studies cited in support of the gateway theory show that people using heroin or crack used marijuana first. Unfortunately, these data tell little about the magnitude of the association between marijuana and hard drugs. Because only the users of hard drugs serve as participants, the data neglect the many, many people who consumed cannabis but no other illicit drugs. The use of marijuana also does not always precede the use of harder drugs, limiting the support for temporal antecedence. That is, some people try crack or ecstasy before they smoke cannabis. These data also fail to isolate marijuana as a cause. The association, temporal antecedence, and isolation criteria are explained in more detail below.

Association

Hume's first criterion for establishing a cause concerns association. Despite popular stereotypes, the association between marijuana and harder drugs is not particularly strong. Many people who abuse hard drugs used marijuana first, but few people who smoke marijuana go on to consume other intoxicants. One study found that 75% of men who used marijuana between 10 and 99 times never used any other illicit drug (Kandel & Davies, 1992). Part of the absence of an association stems from the small number of people who use hard drugs relative to the many who have tried marijuana. Data from the Substance Abuse and Mental Health Services Administration (SAMHSA, 2000) can shed light on this question (see table 3.1).

Most associations are expressed as a correlation coefficient, a number ranging from −1 to +1 that depicts how well two phenomena go together. If everyone who tried marijuana also snorted cocaine, their correlation would be 1.0. Few phenomena go together perfectly, but larger numbers (up to 1.0) mean a stronger relationship. Correlations around .3 frequently receive some attention in the social sciences. For example, Scholastic Aptitude Test math scores correlate approximately .3 with grades in math classes (Gougeon, 1984). Personality measures and alcohol consumption often correlate between .3 and .4 (Earleywine, Finn, & Martin, 1990; Earleywine & Finn, 1991). Correlations smaller than .3 often attract little attention.

Table 3.1. 1999 Drug Use Rates in America (in millions)

	Marijuana	Cocaine	Crack	Heroin
Lifetime	76.4	25.4	5.9	3.0
Past year	19.5	3.7	1.0	0.4
Past month	11.1	1.5	0.4	0.2

Data from Substance Abuse and Mental Health Services Administration (SAMHSA) (2000). Summary of findings from the 1999 National Household Survey on Drug Abuse, Rockville, MD: SAMHSA.

Based on the national survey data, the actual correlation between marijuana and crack cocaine use is .02. This calculation assumes that everyone who used crack cocaine used marijuana first. In addition, this calculation takes into account the many people who use marijuana and never try the harder drug. The correlations between marijuana and heroin are even smaller. (See the final section of this chapter for the computations.) Thus, most correlations that scientists view as important are over 15 times larger than the correlation between marijuana and crack cocaine use.

Standard measures of the correlation between marijuana use and the use of harder drugs are very small. They offer little support for an association between marijuana and other drugs. In an effort to depict the link between marijuana and other drugs in a different way, some authors have turned to another statistical procedure known as conditional probabilities. Conditional probabilities reveal the chances that people will try a harder drug if they have tried marijuana. They are computed by dividing the number of users of the harder drug by the number of marijuana users. If everyone who smoked marijuana tried crack, the conditional probability would be 1.0. If half the marijuana users tried crack, the conditional probability would be .5. If we had 100 people and 30 of them used cannabis, the conditional probability of using cannabis would be .3. If 10 of those 30 went on to try crack, the chance of using cocaine given that cannabis was used first would be one-third.

Again, assume that everyone in the national survey who used a hard drug tried marijuana first. The chance of trying powder cocaine after trying marijuana appears to be relatively high. If all 25 million people who tried cocaine were among the 76 million who had also tried marijuana, 25 million divided by 76 million equals .33. Thus, 1 in 3 people who try marijuana also try cocaine. This number is certainly large enough to warrant concern. Statistics like these motivated drug reformers in the Netherlands to remove criminal penalties for small amounts of marijuana in hope of separating it from the cocaine drug market. They also inspired increased penalties for cannabis in the United States.

The conditional probability linking marijuana to cocaine may be alarming. Nevertheless, the number of marijuana users who continue to use cocaine regularly is markedly smaller. Although 25 million Americans have used cocaine in their lifetimes, fewer than 4 million used it in the past year. Assume that these people use the drug at least once a year, and that they all tried marijuana first. Thus, the chances of marijuana leading to yearly cocaine use are 4 million (the number of yearly cocaine users) divided by 76 million (the number who tried cannabis), or about .05. This means only 1 in 20 people who try marijuana use cocaine once a year or more. Even fewer people used cocaine in the last month (1.5 million). Comparable computations suggest that the chances of trying marijuana and then using cocaine monthly are .02, or about 1 in 50. Thus, less than 2 in 100 marijuana users go on to use cocaine monthly. The probabilities for using crack cocaine and heroin are even lower (see table 3.2).

Thus, the association between marijuana consumption and problem use of other drugs is very small. In fact, studies of the gateway theory do not assess the problems associated with hard drug use; they simply focus on trying other substances. Some studies consider trying a drug a single time as confirmation of use (e.g., Blaze-Temple & Lo, 1992), an approach that has drawn criticism (Zimmer & Morgan, 1997). Trying marijuana is not sufficient to cause the use of harder drugs. Even if the association were markedly larger, this criterion alone does not establish causality. Causal arguments require temporal antecedence and isolation, too.

Table 3.2. Gateway Associations

Of all Americans who have tried marijuana
Few have used marijuana regularly
approximately 1 in 4 used marijuana in the past year (25.5%)
approximately 1 in 7 used marijuana in the past month (14.7%)
Few people who have tried marijuana have used cocaine
approximately 1 in 3 tried cocaine (33.0%)
approximately 1 in 20 used cocaine in the past year (4.8%)
approximately 1 in 50 used cocaine in the past month (2.0%)
Even fewer people who have tried marijuana have used crack
approximately 1 in 13 tried crack (7.7%)
approximately 1 in 100 used crack in the past year (1.3%)
approximately 1 in 200 used crack in the past month (0.5%)
Even fewer people who have tried marijuana have used heroin
approximately 1 in 26 tried heroin (3.9%)
approximately 1 in 200 used heroin in the past year (0.5%)
approximately 1 in 333 used heroin in the past month (0.3%)

These calculations assume that everyone who ever tried hard drugs used marijuana first. Data from Substance Abuse and Mental Health Services Administration (SAMHSA) (2000). Summary of findings from the 1999 National Household Survey on Drug Abuse, Rockville, MD: SAMHSA.

Temporal Antecedence

If marijuana actually caused the consumption of other intoxicants, cannabis consumption must precede the use of hard drugs. Many studies show that hard drug users smoke marijuana first. The stereotyped progression of the use of drugs probably begins with caffeine, but no data address this drug's potential as a gateway. Most research suggests that adolescents first use alcohol or nicotine. Some authors argue that cigarettes serve as the actual gateway to drug problems (Kandel et al., 1992; Labouvie, Bates, & Pandina, 1997). A subset of the people who drink alcohol and smoke cigarettes subsequently use marijuana. A subset of those who try marijuana then use harder drugs like cocaine, crack, and heroin. Some researchers report that over 90% of people who try hard drugs tried marijuana first, suggesting temporal antecedence. They also ate white sugar, breathed air, and attended grade school. Thus, although hard drug users smoke marijuana before turning to other substances, this fact alone does not prove causality.

In addition, the order of drug use is not always perfectly consistent with the idea that marijuana is the gateway. Many hard drug users do not start with marijuana. Allen Ginsberg, the celebrated "beat" poet, serves as one notable exception. He injected heroin before smoking cannabis (Ginsberg, 1966). Large samples with thousands of respondents show a range of hard drug users who did not try marijuana first. One study found that as few as 1% of people who used hard drugs had not tried marijuana (Donovan & Jessor, 1983). In contrast, 15% of another sample of heavy drug users started with cocaine or intravenous drugs before smoking marijuana (Golub & Johnson, 1994). A study of Australian youth found that 29% of those who had used amphetamine, LSD, cocaine, or heroin had not used marijuana first (Blaze-Temple & Lo, 1992). Other research found 39% of a sample used hard drugs before trying cannabis (Mackesy-Amiti, Fendrich, & Goldstein, 1997). Thus, temporal antecedence applies in some cases of hard drug use, but not all. Even a perfect ordering with marijuana preceding all other drug use does not prove causality without isolation.

Isolation: Independent Processes or Problem Behavior?

Hume's last criterion for causality concerns isolation. If marijuana actually causes the use of hard drugs, other explanations should not account for any association between them. At least two alternative explanations have some empirical support. One concerns the idea that the initiation of each drug arises from its own individual process. Miller (1994) calls this approach the statistical independence hypothesis. This hypothesis states that using one drug stems from its own availability, expectancies, and motivations that are separate from those related to another drug. Thus, the initiation of caffeine may stem from one process; the initiation of heroin may stem from another. The other idea that may account for stages of drug use focuses on the abuse of any and all substances as part of a cluster of larger problem behaviors. This problem behavior theory views substance abuse, unsafe sex, crime, and delinquency as all part of the same underlying trouble.

The rationale for individual processes follows statistical logic. People who participate in rare events likely engage in popular activities first. For example, more people view television than skydive. Thus, we would expect that most skydivers watched TV before they leaped from a plane. This fact need not mean that television causes skydiving. The two acts probably arise from independent, individual processes. Yet the most common one occurs first simply because it is more common. Comparable logic applies to drug consumption. People who use drugs will likely begin with those that are most common. Thus, individuals may use marijuana before cocaine because marijuana is more prevalent in our culture. Yet this fact need not mean that marijuana caused cocaine consumption. A massive study of four national samples including over 6,000 participants suggests that a large portion of the appearance of stages of drug use can be accounted for by statistically independent processes (Miller, 1994). This model does not account for all the data, but independent processes clearly contribute to the progression of drug use.

An alternative way to test the statistical independence hypothesis might examine neighborhoods where crack cocaine is more available than marijuana. If most drug users in such neighborhoods smoked crack before cannabis, the role of availability might receive some support. Under these circumstances, few could conclude that crack is a gateway leading to cannabis. Instead, people use the most available drug first and less available drugs later or not at all.

Another alternative explanation of drug sequencing concerns problem behavior theory. According to this theory, a small group of adolescents engage in a cluster of actions that all may lead to negative consequences (Jessor & Jessor, 1977). These problem behaviors include drug consumption, poor school performance, unsafe sex, and criminal activities. According to this theory, the association between cannabis use and the consumption of other drugs does not arise because marijuana causes problems with other substances. Instead, both marijuana and the use of other intoxicants arise from the underlying problem orientation in a subset of individuals.

Many studies reveal strong correlations among the use of different drugs (Earleywine & Newcomb, 1997). Several potentially dangerous actions also correlate with drug use (Jessor, 1998). Miller (1994) analyzed data from four national surveys including more than 6,000 participants and found that problem behavior theory may account for the appearance of stage-like progressions in substance use. He found a large subgroup who used many drugs, and another set of people who used no drugs at all. These results are consistent with the idea that an underlying "problem-proneness" may account for links between marijuana and other drugs. Essentially, cannabis does not cause cocaine consumption, but a subset of people who like marijuana also like cocaine.

In addition to statistical independence and problem behavior theory, a third set of findings also supports arguments against marijuana as an isolated cause of hard drug use. Studies that show personality traits correlate with the use of multiple substances may mean that a personality characteristic led to both marijuana use and hard drug use. These data suggest that the same personality traits that can lead to smoking cannabis can also lead to snorting cocaine. Thus, the marijuana may not cause the use of the other drugs; both stem from the same underlying characteristic. Although evidence for an addictive personality is clearly limited (Nathan, 1988), people who report strong desires for thrill, adventures, and sensations often use a greater variety and amount of drugs (Simon, Stacy, Sussman, & Dent, 1994).

These findings support the idea that marijuana cannot be isolated as the cause of the use of hard drugs. Simple exposure to cannabis is not strongly associated with the use of other intoxicants. Hard drug users do not always use marijuana first. Causes other than marijuana also lead to the consumption of heroin or cocaine. Nevertheless, some authors argue that marijuana may still contribute to the use of harder drugs, even if it is not a unitary cause. They assert that even if cannabis does not qualify as the cause of other drug problems, it facilitates the use of more substances, increasing the likelihood of trouble.

Marijuana as Contributor Rather than Cause

Few interesting behaviors, including drug use, arise from a single process. Data do not support the idea that smoking marijuana causes the abuse of other substances. Nevertheless, perhaps marijuana "contributes" to the use of other drugs, even if it does not serve as the sole cause. This contributing process could work in several ways. For example, smoking marijuana may lead people to think of themselves as illicit drug users, making hard drug consumption more likely. Another pathway may arise when purchasing marijuana exposes people to the market for other drugs. In addition, marijuana intoxication may limit a person's ability to refuse harder drugs when they are offered. These factors may interact to contribute to the transition from cannabis consumption to hard drug use. Each of these paths is discussed in the following paragraphs.

One potential path for marijuana's impact on substance abuse concerns an individual's identity as an illicit drug user. Despite rampant consumption of caffeine and nicotine, few of us see ourselves as drug users. Most people who have never consumed an illicit drug would claim that they have no intention of snorting cocaine or smoking crack. Yet people might alter their impressions of themselves after smoking marijuana. With continued consumption of cannabis, people may see themselves as illicit drug users. After establishing this identity, their chances of trying other drugs may increase. Thus, people who may have had no intention of using cocaine or crack before they smoked marijuana may consider consuming these drugs after a period of cannabis use. Researchers have yet to examine this potential path, but studies along these lines may prove fruitful. Longitudinal studies might assess each individual's identity as a drug user over time. Some of those who try marijuana may adopt this identity; others may not. If those who smoke cannabis and subsequently consider themselves illicit drug users make the transition to hard drugs, the data would support this theory.

Another pathway may involve exposure to the illicit drug market. Suppliers of cannabis may also sell harder drugs, exposing marijuana purchasers to cocaine or heroin. A supplier may have a set of marketing strategies for these harder drugs, including strong personal testimony about their quality. Each purchase of marijuana may expose people to sales pitches for more harmful drugs, perhaps increasing the likelihood of eventually trying them. Few studies address this course, but it has a certain intuitive appeal. This line of reasoning motivated marijuana decriminalization in the Netherlands in an attempt to separate it from the hard drug market. Some studies suggest that this move decreased the consumption of hard drugs (see chapter 10). Other studies might include interviewing hard drug users to see if they first obtained these drugs from the same person who supplied their cannabis.

Finally, perhaps marijuana users would be more likely to try other drugs during intoxication. People often use combinations of different substances (Earleywine & Newcomb, 1997). Yet no data address if people first try cocaine or heroin while experiencing cannabis's effects. The cognitive impairments associated with marijuana intoxication might decrease an individual's ability to resist using other drugs. A couple of experiments might shed light on this phenomenon. Researchers might provide access to cocaine or heroin after injecting THC into rodents to see if they are more likely to consume the new drugs during intoxication. Studies of human reactions might rely on self-reported willingness to try other drugs after smoking marijuana. People may claim to be more likely to ingest a new substance after consuming cannabis than they would after smoking a placebo. These data would support the idea that marijuana intoxication alters the probability of consuming hard drugs.

Preventing Substance Abuse

If marijuana consumption led directly to the abuse of hard drugs, preventing substance abuse would be simpler. In fact, drug problems arise from complex interactions of multiple factors. Preventing addiction and other aspects of substance abuse proves difficult. Many attempts have been relatively unsuccessful, including scare tactics, basic drug education, the Drug Abuse Resistance Education (DARE) program, and enhancing self-esteem. Other programs show more promise, including interactive sessions that teach techniques for combating peer pressure to use drugs. Social influence programs combine multiple strategies and have proven effective at decreasing drug use (Sussman, Dent, Stacy, & Craig, 1998). Unfortunately, no program has consistently eliminated drug abuse in all participants, and the impact of prevention efforts can often dissipate quickly (Shope, Copeland, Kamp, & Lang, 1998). Deterring drug problems remains an ongoing challenge.

Efforts to prevent drug-induced harm are likely as old as the drugs themselves. Some relied on outrageous punishments, such as the Czar of Russia in 1634 who slit the nostrils of soldiers found smoking tobacco (Maisto, Galizio, & Connors, 1995). Some current attempts to minimize drug problems are less punitive and stem from programs developed in the late 1960s and early 1970s. At that time, the sudden widespread use of marijuana, barbiturates, amphetamines, and hallucinogens made many adults panic. They quickly designed programs to arouse fear of the adverse consequences of drugs. Unfortunately, some of the information appeared exaggerated or flagrantly wrong. These errors eroded the credibility of the presentations, leading young people to doubt the entire content of the program. Thus, the scare tactics proved ineffective in preventing the use of alcohol and other drugs (Goldberg, Bents, Bosworth, Trevistan, & Elliot, 1991; Powers-Lagac, 1991). Paradoxically, some rebellious adolescents viewed the threat of negative consequences as a challenge. They may have initiated drug use in an effort to show courage. Thus, sometimes scare tactics actually backfire and lead to more drug use.

Perhaps in reaction to the failure of scare tactics, some prevention experts adopted an opposite approach. They supplied objective information about illicit substances and their consequences without exaggeration or drama. Designers of these standard drug education programs hoped that this tactic would enhance their credibility and help students make informed decisions. Ideally, the candid assessment of the negative consequences would lead students to stay away from drugs. Participants in these programs invariably learned a lot about different substances. Most studies showed that students exposed to this information learned a great deal about drugs of abuse and their potential adverse consequences. Despite increasing knowledge, this approach often had no impact on drug use. Some participants simply became more knowledgeable users. These programs occasionally had the undesired effect of increasing curiosity about illicit substances (Schinke, Botvin, & Orlandi, 1991). Nevertheless, many programs still include objective information to boost credibility.

The DARE program remains one of the best-known, school-based attempts to minimize substance abuse. In 1983, the Los Angeles Police Department and Los Angeles United School District collaborated to bring law enforcement officers into elementary schools to teach general information, drug refusal, and self-management. The project became incredibly popular, spreading to all 50 states and many foreign countries. Students and police officers reported enjoying the program; parents and other members of the community felt good about it, too (Levinthal, 1999). Despite this popularity, DARE has no meaningful impact on drug use. At least eight studies show little difference between students who participated in the program and those who did not (Ennett, Tobler, Ringwalt, & Flewelling, 1994). Research on over 1,000 participants showed that the program never created a better outcome than standard drug education (Lynam et al., 1999). Perhaps expectations for the program were too high. For example, some data looked at students 10 years after the program. Few psychological interventions have such a dramatic impact that their positive effects remain for 10 years. Nevertheless, given the expense of the project and its limited impact, children may benefit more from spending their time in pursuit of a classical education than listening to a police officer tell them to abstain.

Another approach designed to decrease drug abuse focused on enhancing self-esteem. Self-esteem correlates with drug consumption in some studies (Hoefler et al., 1999), suggesting that an improved sense of pride and efficacy might prevent substance abuse. Prevention programs of this ilk boosted perceptions of self-worth quite consistently. Paradoxically, some studies found higher self-esteem was associated with more drug use (Newcomb, McCarthy, & Bentler, 1989; Stein, Newcomb, & Bentler, 1996). Perhaps individuals felt so confident that they believed they could handle drugs without developing problems. Although improving positive feelings about individual successes and personal qualities may increase happiness, these strategies do not prevent drug consumption (Donaldson et al., 1996).

Although the strategies above have had little impact on substance abuse, others show some promise. Social influence approaches to preventing substance abuse have met with considerable success. These programs combine several different techniques to inoculate people against social pressures to use drugs. The most successful programs use an interactive style, encouraging contributions and involvement from participants. The programs begin with basic information on the physical consequences of drug use, with emphasis on long-term and short-term negative effects. They then review techniques for making decisions about drugs that will minimize negative consequences. These techniques usually require an assessment of the pros and cons of consumption. Participants then make a public commitment to a drug-free lifestyle, often by standing in front of other participants to declare their intention. These activities have helped alter attitudes about drugs in other interventions. In addition to these activities, social influence programs add unique components to improve outcomes.

The unique components of social influence programs include increasing awareness about the actual number of drug users, developing skills for refusing drugs when offered, and combating indirect social pressures to use drugs. Perceptions of the number of users often lead to a skewed sense of drug consumption. People often hold the faulty idea that drug use is extremely common, and may feel that they should use drugs in order to obey this norm. Drug users often assume that everyone else uses drugs, too. Perceptions of norms for behavior guide actions in many domains. Objective information about drug consumption often surprises participants, who frequently overestimate the incidence of use (Perkins, Meilman, Leichliter, Cashin, & Presley, 1999). Accurate assessments of the number of users may minimize pressures related to the idea that everyone else consumes illicit drugs.

Drug refusal skills focus on resisting peer pressure to consume drugs. These include more than Nancy Reagan's simplistic recommendations to "just say no." Students engage in role plays designed to enhance their ability to decline drugs whenever they are offered. Strategies include keeping their refusals direct, suggesting alternative activities, avoiding situations where drugs are prevalent, and walking away if pressures feel threatening. Increasing awareness of indirect pressures to use drugs often focuses on the inaccuracy of glamorized media portrayals of substance use. Indirect pressures also may arise when social models, including parents, older siblings, and peers use drugs. Increasing awareness about these pressures may help inoculate against drug use (Donaldson et al., 1996). The majority of programs that employ this social influence approach have had positive effects (Hansen, 1992). Implementing them in more settings with detailed follow-up research can help minimize drug problems in ways that simplistic assumptions about marijuana as a gateway cannot.

Conclusions

The idea that marijuana serves as a gateway or stepping-stone to the consumption of harder drugs with worse negative consequences has generated considerable interest. There is no evidence that cannabis creates physiological changes that increase the desire for drugs. The idea that marijuana causes subsequent drug use also appears unfounded. Causes require association, temporal antecedence, and isolation. Evidence for the association between marijuana and other drugs remains limited. Data do reveal that the majority of cocaine and heroin users consumed cannabis first. Nevertheless, only a minority of marijuana smokers try cocaine, crack, or heroin. Only a few people become regular users of these intoxicants. In addition, marijuana does not precede the use of hard drugs in all cases. Finally, correlations between marijuana smoking, hard drug consumption, and other problem behaviors suggest that one drug may not lead to another so much as all use of illicit substances reflects an underlying deviance or personality characteristic. Thus, prevention of drug problems requires more than staying away from cannabis.

Many programs designed to minimize substance abuse have met with only limited success. The scare tactics of the 1960s and 1970s had little impact on drug use. The DARE program, though popular, does little to prevent the use of illicit substances. The enhancement of self-esteem also does not prevent drug problems. One series of studies suggests that programs designed to minimize the impact of social influences to use drugs holds considerable promise. By providing valid information on the relative infrequency of drug use and valuable coaching on ways to resist pressures to use intoxicants, these programs can help decrease the incidence of abuse and dependence.

Comment on the Computation of Correlations

A few adventurous souls may wish to know the exact correlation between marijuana use and the use of other drugs. A Pearson product-moment correlation can be computed from a 2×2 table using the cross products and the marginals, in a manner that is easier done than said. Cross products are computed by multiplying the numbers along the diagonal. Marginals are the totals across each row and column. The correlation (R) equals the difference in the cross products divided by the square root of the product of all four marginals (Rosenthal & Rosnow, 1991). Using the data on drug use from 1999, and an assumption of 221 million adult Americans, we can compute the correlation between marijuana and crack cocaine use. Assume absolutely everyone who tried

Table 3.3. Millions of Americans Who Have Tried Marijuana, Crack Cocaine, Both, or Neither (1999)

	Tried marijuana		
Tried crack	Yes	No	Total
Yes	6	0	6
No	70	145	215
Total	76	145	

crack tried marijuana previously. The number of people (in millions), who tried marijuana, crack cocaine, both, or neither, appears in table 3.3.

Thus, of 221 million adult Americans, 76 million tried marijuana and 145 million did not. Assume that the 6 million Americans who tried crack cocaine also tried marijuana. Thus, 70 million people who tried marijuana never tried crack. The correlation from this table is computed by starting with the difference in the cross products. Moving from the upper left to lower right on the diagonal, we have 6 million $\times$ 145 million = 8.7×10^{13}. The upper right times the lower left is 0 $\times$ 70 million = 0. The difference between these two equals 8.7×10^{13}. We divide this number by the square root of the product of the four marginals. SQRT (76 million $\times$ 145 million $\times$ 6 million $\times$ 215 million)= 3.77×10^{15}. We divide 8.7×10^{13} by this number and get a correlation of .02. The numbers for heroin are even smaller because so few people have ever tried it. Thus, the correlation between marijuana consumption and the regular use of these harder drugs is negligible.

Marijuana's Impact on Thought and Memory

People think and remember differently during cannabis intoxication. Chronic consumption of the drug might change aspects of cognition, too. The deficits associated with intoxication are relatively specific. People who are high show obvious problems concentrating, attending to details, focusing on goals, performing two actions simultaneously, and learning new, complex information. These problems grow worse with higher doses of the drug and more complicated tasks. In some studies, marijuana intoxication impairs the ability to react quickly, show restraint, and persist with dull exercises. Intoxicated people should probably avoid any task that requires fast reflexes or sustained attention. Many other facets of thought remain intact after smoking cannabis, including the ability to learn simple tasks and remember material mastered prior to using the drug.

Intoxication clearly alters some aspects of thought and memory. In addition, chronic exposure to cannabis may change cognition. Some studies reveal cognitive problems associated with long-term use of cannabis, but many others do not. Some of the first research on the effects of chronic marijuana exposure did not reveal much impact of the drug. Only a few studies revealed altered thinking and memory in people with many years of daily use, and other research found that chronic users performed as well as nonusers on plenty of measures. These studies suggested that chronic smoking of marijuana likely does not produce major changes in general cognitive abilities like intelligence, memory, and the ability to learn.

This absence of gross impairments is reassuring, but research employing more sensitive measures reveals subtle deficits in users with prolonged, frequent exposure to the drug. These people appear to process information less quickly and efficiently. Chronic, heavy marijuana smokers also show deviant brain waves when performing certain complex tasks. The practical implications of these effects remain unclear. Critics of this work emphasize many methodological flaws. Nevertheless, research still suggests that chronic use of marijuana leads to subtle problems in complex tasks. Each of these effects and the relevant research issues appear below.

Overview of Acute Effects

Cannabis intoxication alters thoughts. The exact extent of the alteration varies with dosage, setting, experience, and other factors. Higher doses in novel, laboratory settings may produce dramatic impairment on some tasks, particularly in people who have little experience with the drug. Cannabis has a varied impact on different measures. Generally, marijuana does not alter performance on easy tasks but impairs complex ones. For example, simple learning tasks like memorizing pairs of words show little change during intoxication. Memory for material learned prior to intoxication also remains intact while participants are high. For example, participants who memorize a list of words before they smoke cannabis can remember the list during intoxication.

Laboratory evidence is inconsistent for marijuana intoxication's effect on some other tasks. A few studies suggest intoxication hurts performance, but others do not. For example, cannabis makes simple reaction times longer in some studies but not others. Intoxicated people can have trouble solving problems in new ways, but not in all studies. Some research suggests that people cannot pay attention for long durations after smoking marijuana; other studies show no problems on this sort of task. Another set of skills clearly suffers after smoking marijuana. Intoxication consistently impairs time perception, reading difficult material aloud, mental arithmetic, complex reaction time, and certain aspects of memory and perception. A summary of the acute effects on cognitive tasks follows:

Probably unaffected
easy learning
remote memory

Possibly affected
simple reaction time
disinhibition
vigilance

Probably affected
perception
reading aloud
arithmetic
complex reaction time
recall
intrusions in recognition memory

Acute Effects

Probably Unaffected Tasks

Proving that a drug has no effect is difficult. A single study may not have enough participants to reveal cannabis-induced changes, the dosage of marijuana may be too small to achieve an effect, or the tests may be too easy. Nevertheless, multiple studies with proper doses and adequately large samples suggest that cannabis intoxication does not impair the ability to learn simple tasks or remember material mastered before intoxication.

SIMPLE LEARNING—PAIRED ASSOCIATES One version of paired associates learning requires reading pairs of words. After a delay, participants view one word and attempt to recall the associated word from the previously learned pair. For example, a list might include the word "baby" paired with "red." The experimenter might then present the word "baby." The participant should then respond with "red." Marijuana intoxication does not appear to alter performance on this task. Intoxicated individuals recall the appropriate words as often as people who smoked placebo pot, suggesting that the drug does not impair simple learning (e.g., Chait & Pierri, 1992; Hooker & Jones, 1987).

REMOTE MEMORY Although cannabis intoxication is notorious for its impact on memory, some domains of recall show little impairment. Remote memory, which concerns the ability to retrieve material already learned, does not suffer during intoxication. Most studies of remote memory during intoxication ask participants to recall words that they learned prior to ingesting THC. In other studies, intoxicated people list all the words they know that begin with a specific letter. One research team asked participants to remember TV shows that ran for only a single season many years previously (Wetzel, Janowsky, & Clopton, 1982). In each case, participants knew the information before they smoked marijuana in the laboratory. The drug did not impair these remote memory tasks (Chait & Pierri, 1992). Thus, intoxicated people can likely remember material they learned before smoking marijuana—from their first grade teacher's name to recently learned lists of words. In fact, some users report that during intoxication they are more likely to spontaneously remember remote events from their past that they had not recalled in years (Tart, 1971). This claim has not been tested empirically but seems consistent with the absence of an impact on remote memory.

Possibly Affected Tasks

Marijuana has had inconsistent or limited impact on simple reaction time, disinhibition, and vigilance. Thus, it is unclear if cannabis intoxication prevents people from responding quickly, inhibiting automatic reactions, and persisting on long, tedious assignments. Some studies have used small samples, making small effects difficult to detect. A few studies used small doses of the drug. These did not impair performance, but larger doses might have. Further work in these areas may help clarify the drug's effect on these abilities.

SIMPLE REACTION TIME Researchers assess reaction time by asking participants to press a switch as quickly as possible after hearing a tone or seeing a light. This task usually requires quick thinking and quick reflexes, but does not demand any complex decisions. Marijuana intoxication had small but statistically significant effects in some studies (Borg, Gershon, & Alpert, 1975; Dornbush, Fink, & Freedman, 1971), but no effect in others (Braden, Stillman, & Wyatt, 1974; Evans, Martz, Rodda, Lemberger, & Forney, 1976). The studies that found no impact of marijuana used comparable dosages, suggesting that the absence of an effect did not arise from too little of the drug. These studies also examined as many participants as those that found marijuana impaired reaction time, suggesting that the absence of an effect did not arise from failing to test enough people. Thus, the lack of an effect does not seem to stem from inadequate methods. Only further work can reveal the exact impact of intoxication on simple reaction times.

The potential effect of marijuana on reaction time creates difficulties for other research on cognition. Many tests of thinking rely on speeded performance. For example, measures of the efficiency of thought often look at the time required to press a correct button. Any test that requires quick responding might show marijuana-induced impairments simply because of the drug's impact on reaction time. Nevertheless, given the small and inconsistent effects, researchers generally assume that the impact of marijuana on other tasks probably does not stem from a simple problem with reaction time (Chait & Pierri, 1992).

DISINHIBITION To the disappointment of many, much of adult life requires consistent restraint. The inability to inhibit can lead to a broad array of problematic actions, including troublesome overeating, frequent intoxication, ill-advised sexual encounters, and violent outbursts. These behaviors can have negative repercussions that range from embarrassment and illness to unemployment or imprisonment. Although researchers cannot measure all of these consequences of disinhibition in the laboratory, a few creative tasks have been developed that seem to relate to the general ability. The impact of marijuana on these tasks has been mixed.

A few studies have examined cannabis's effect on the Stroop task. This task requires looking at the names of colors (e.g., red, green, blue) printed in colored ink. The color of the ink may not match the word. For example, the word "BLUE" might appear in red ink. Instead of reading the words, participants must name the color of the ink. This task is difficult because people have much more practice reading words aloud than naming colors. The correct response to the word "BLUE" printed in red ink is to say the word "RED," the ink's color. This response requires inhibiting the dominant reaction, to read the word "BLUE."

People take longer to name the colors of these words than they take to name the colors of a series of letters that do not spell words. The fact that the words are the names of colors interferes with naming the color of the ink. Researchers generally interpret this task as a measure of disinhibition. Participants must inhibit the dominant, reading response to perform the less-practiced naming of colors. People who have problems inhibiting themselves often do more poorly on this task. For example, children with attention deficit disorder and adults with alcoholism show problems with this task (Gorenstein, 1987; Gorenstein, Mammato, & Sandy, 1989). This interference with color naming has increased during marijuana intoxication in some studies (e.g., Hooker & Jones, 1987), but not others (Evans et al., 1973; Chait & Pierri, 1992). Notably, the studies that reveal no effect often have smaller samples, which decreased the chances of revealing a marijuana-induced deficit. The positive results suggest that marijuana impairs the ability to inhibit responses. People who have recently smoked cannabis may do a poor job of inhibiting in other domains, too. They may overeat, have ill-advised sexual encounters, blurt inappropriate words, and fail to resist other temptations.

VIGILANCE Studies of vigilance or sustained attention usually require extended periods of concentration on an exceedingly simple, dull task. These studies prove important because many users claim that their work on repetitive chores improves during intoxication (Carter, 1980). One task, the continuous performance test, requires watching a series of digits show up on a screen and hitting a key when the number 8 appears. It is barely more interesting than watching water evaporate. Marijuana does not impair or improve performance on this task (e.g., Vachon, Sulkowski, & Rich, 1974). This measure of vigilance may be too easy to reveal any changes. Studies only required up to 7 minutes of performance; perhaps deficits or improvements may appear after longer durations.

In another typical study of vigilance, participants watched a circle of neon bulbs. The bulbs lit, one at a time, in succession. Occasionally, one light in the circle was skipped. Participants were required to press a button any time a light was skipped. They performed this task for an hour, which must have been about as interesting as a documentary on how to make mud. Perhaps this task is comparable to certain forms of employment. After smoking marijuana, people tended to miss the skipped lights more often. Both the intoxicated and the unintoxicated people had more misses as time went on, but the rate of decline was worse after smoking marijuana. This result suggests that marijuana intoxication may decrease vigilance on long, dull tasks, contradicting anecdotal reports of improvement (Sharma & Moskowitz, 1974). Further work might examine if people develop tolerance to this effect of marijuana. Financial incentives for better performance would probably decrease errors during intoxication, too. These investigations might reveal the seriousness of marijuana's impact on vigilance.

Probably Affected Tasks

Intoxication likely impairs perception, reading aloud, arithmetic, complex reaction time, and certain aspects of memory. These effects appeared at standard doses even in small samples.

PERCEPTION Marijuana intoxication alters the senses. People report changes in taste, touch, smell, sight, and hearing. Laboratory data confirm some of these reports. At least 10 studies show that smoking marijuana alters time perception. Apparently, intoxicated individuals experience time as passing more slowly. Subjective reports suggest that events seem to take longer, as if a few seconds feel like a minute. For example, during intoxication, a single recording album may seem to take hours to play (Tart, 1971). Laboratory studies confirm that intoxicated individuals perceive brief intervals as markedly longer than they are in reality. After smoking marijuana, people asked to wait 30 seconds think that they have waited more than 30 seconds. Intoxicated people asked to signal when they think 30 seconds have passed often respond after only 20 seconds (Chait & Pierri, 1992). Perceptions of space also change. After smoking marijuana, people report that the distance between objects seems to increase (Tart, 1971). At least one study using a driving simulator confirms that marijuana intoxication alters the perception of distance. Intoxicated people driving a simulator tended to overestimate how far they traveled (Bech, Rafaelsen, & Rafaelsen, 1973).

Cannabis intoxication also appears to alter vision. After smoking marijuana, people do not distinguish colors well. They show problems discriminating among shades of blue (Adams, Brown, Haegerstrom-Portnoy, & Flom, 1976). They also process cues for three dimensions differently, making it more difficult for them to enjoy certain illusions of depth (Emrich et al., 1991). Intoxicated individuals appear less able to identify figures hidden within pictures, too (Pearl, Domino, & Rennick, 1973). Because accurate processing of information requires accurate initial perceptions, these impairments may underlie a number of cognitive distortions associated with marijuana intoxication. For example, any test that requires quick responding to specific colors will show impairment simply because colors are not perceived correctly. Any further processing of the colors may actually remain intact, but because the initial input is faulty, all subsequent processing appears incorrect. For example, this impaired perception of colors may have contributed to altered performance on the Stroop color-naming task.

READING ABILITY In a series of studies, participants read unfamiliar passages of difficult text forward or backward while hearing their own voices though earphones. The researchers made the task more difficult by employing a manipulation known as delayed auditory feedback. As they read, the participants heard their own words played a quarter of a second after they said them. This manipulation parallels some of the annoying qualities of being mocked. The texts employed in these studies were also quite complex. Some studies asked participants to read a section of Aristotle's work aloud under these conditions (Manno, Kiplinger, Haine, Bennett, & Forney, 1970). Marijuana intoxication consistently impaired reading aloud under these circumstances. Intoxicated individuals took longer to read the text and made more mistakes (Chait & Pierri, 1992). These data suggest that cognitive abilities are not at their peak during intoxication. Their practical implications remain less clear. Anyone planning a public reading of *The Nichomachean Ethics* should probably avoid cannabis.

ARITHMETIC Marijuana impairs mathematical performance. In a dozen studies intoxicated participants tried to add or subtract a series of digits. Most research showed marijuana-induced deficits, particularly for the more complex tasks. People asked to count backward from 100 by 7s tend to perform worse after smoking marijuana. People who had to perform more complex addition and subtraction problems also showed deficits when high (e.g., Casswell, 1975; Casswell & Marks, 1973). These data further support the idea that cannabis intoxication decreases mental abilities, particularly the attention associated with computation. People may develop tolerance to these effects, but no studies have addressed this question. Obviously, anyone who has smoked marijuana and must compute a tip or balance a checkbook should wait until intoxication wears off or rely on a sober friend.

COMPLEX REACTION TIME Although simple reaction times do not always appear impaired during marijuana intoxication, as tasks grow more complicated, performance declines. Complex reaction time tasks usually require pressing different buttons in response to different events. For example, experimenters might ask participants to press one button in response to a green light, and another in response to a red light. The tasks reveal different marijuana-induced impairments. Sometimes intoxicated participants press the wrong button more often (Low, Klonoff, & Marcus, 1973), though not always (Peeke, Jones, & Stone, 1976). Their reaction times also increase. Some studies show small increases in reaction time, roughly 10% (Borg et al., 1975), but others show intoxicated participants take up to 50% longer to make the correct response (Block & Wittenborn, 1986). Generally, the more complex the task, the worse the cannabis-induced impairment (Clark & Nakashima, 1968; Chait & Pierri, 1992). People do appear to develop tolerance to this effect with repeated practice while intoxicated (Peeke et al., 1976).

MEMORY Although memory for material learned prior to intoxication often remains intact after smoking marijuana, other aspects of memory decline dramatically. Memory problems associated with intoxication usually appear when researchers present a series of words to people after they have consumed cannabis. Participants then wait briefly and view a second list of words. Some of the words on this second list appeared on the first; participants then guess which ones. The ability to recognize the correct words is known as recognition memory. Intoxicated participants are very good at identifying the words they saw previously (Miller, Cornett, & Wikler, 1979; Miller et al., 1977). Nevertheless, they also claim to recognize words that actually were not on the original list (Dornbush, 1974). These mistakes are known as memory intrusions. Thus, aspects of recognition memory may suffer during intoxication.

These recognition memory problems suggest that marijuana may impair the ability to separate relevant and irrelevant stimuli. An irrelevant word, one that did not appear on the previous list, seems as familiar as relevant words that did appear on a previous list. These results for words do not generalize to all practical aspects of memory. For example, eyewitness testimony does not appear impaired. Marijuana intoxication had little negative impact on the recognition of important information relevant to an event (Yuille, Tollestrup, Marxsen, Porter, & Herve-Hugues, 1998). Perhaps cannabis has less impact on memory for meaningful events like those important to eyewitness testimony, and more impact on memory for meaningless stimuli like lists of words.

Another type of memory, free recall, shows definite impairment during intoxication. In these studies, participants usually write down as many words as they can from a previously presented list. Thus, they must generate the words rather than simply recognize if a given word appeared before. This task proves more difficult than the recognition memory exercise. Intoxicated people invariably remember fewer words (e.g., Dornbush, et al., 1971). They also tend to include words that were never on the list (Miller & Cornett, 1978). These memory intrusions are quite common after smoking cannabis (Chait & Pierri, 1992). They support the idea that marijuana intoxication impairs the ability to separate relevant from irrelevant stimuli.

Summary of Acute Effects

In general, marijuana intoxication has little impact on learning simple tasks or remembering information mastered prior to ingesting the drug. Intoxicated people can probably still become proficient at new, easy skills. They can probably recall events that occurred before they smoked marijuana, including important information from the distant past. The drug has produced inconsistent effects on simple reaction time, disinhibition, and vigilance. Sometimes the drug slows reaction time, makes people unable to restrain their first impulse, and prevents them from sticking with long, dull tasks. Marijuana intoxication clearly impairs aspects of memory, perception, reading, arithmetic, and complex reaction time. After smoking cannabis, people cannot memorize new lists of words, distinguish among similar colors, read complicated passages aloud, subtract strings of numbers, or respond quickly to different lights by pressing different buttons. Any tasks like these that require elaborate, precise, or rapid thinking should probably not be performed during intoxication.

Effects of Chronic Marijuana Consumption

Acute marijuana intoxication alters thought and memory. Many researchers have investigated the effects of chronic use of marijuana on these cognitive functions, too. If a single dose of the drug leads to impairments, perhaps consistent use would lead to comparable deficits, even after intoxication has worn off. Most human studies compare people who smoked marijuana daily for many years to others who report never using the drug. Although much of this work reveals no gross impairment in chronic users, some studies report lower test scores or deviant brain waves in those who smoke daily for extended periods. Interpretations and critiques of these studies vary with their results. Studies that find that chronic smokers perform as well as others are often criticized for having inadequate samples or insensitive tests. Critics of studies that reveal deficits in chronic smokers also focus on the people studied and the tests employed. A review of these critiques appears first, followed by detailed summaries of the relevant studies. In general, despite the many critiques of the research so far, chronic marijuana consumption does not appear to create gross neuropsychological impairments. Regular use for many users does, however, lead to deficits on highly sensitive tests.

Overview of Critiques of Studies That Reveal Few or No Deficits

Many researchers report no differences between people who never use drugs and unintoxicated, chronic users of cannabis. These results suggest that chronic marijuana exposure has little permanent impact on thought and memory. Critics of these studies emphasize that genuine differences may exist, but they failed to appear for several reasons. These reasons relate to both the participants studied and the tests employed. The issues related to participants concern samples that are too small to reveal differences, biased sampling that includes only the most competent marijuana smokers, and contaminated group membership that allows smokers to claim they have not used the drug. Issues related to the tests usually concern their lack of sensitivity to subtle cognitive problems.

Critiques

INSUFFICIENT SAMPLE SIZES The first critique of studies that reveal no changes in cognitive function in chronic marijuana smokers concerns the number of people studied. Some research that fails to show marijuana-related cognitive deficits may not employ enough participants. Research cannot reveal effects without a sufficient number of data points. Some investigators who report no effect of chronic use assessed as few as 10 people (e.g., Schaeffer, Andrysiak, & Ungerleider, 1981). Detecting large differences between users and nonusers likely requires at least 25 from each group (Cohen, 1990). The obvious solution to this problem is to study more people.

BIASED SAMPLING The second critique concerns the characteristics of the participants. First, some investigators of marijuana's effects sample from college or medical students. Because everyone in the sample has to meet certain entrance requirements, only those who have not experienced extreme negative consequences from the drug ever appear in the study. These select groups may give the impression that marijuana causes no harm. In fact, those who were harmed by the drug may never participate in the research because they are not enrolled in college or medical school. Even studies that do not focus on students may fail to sample from the people who are most impaired from marijuana. Volunteers for these studies may not be the most troubled users. Volunteers for research often differ from people who are unwilling to participate, particularly in studies related to drugs (Strohmetz, Alterman, & Walter, 1990). Individuals who experience genuine problems may not prove particularly eager to perform tasks in a laboratory, even for pay. Perhaps many studies sampled only those chronic users who were not experiencing negative consequences. Thus, the representativeness of these samples remains unknown.

This critique is difficult to combat. Many investigators emphasize that their participants have extensive use of the drug for long periods. Some studies employ participants who smoked marijuana daily for over 10 years. Thus, they qualify as the exact people appropriate for study. Another strategy for reaching the most impaired users requires visiting them in their homes. This way, people who may show more impairment or little motivation to visit a laboratory will still appear in the study. Given the illegal status of the drug, this approach may prove quite cumbersome. Nevertheless, Bowman and Pihl (1973) visited some participants in their homes in Jamaica and still found no cannabis-related deficits on a number of cognitive tasks.

CONTAMINATED GROUP MEMBERSHIP Another critique of studies that fail to find differences concerns the validity of the reports of use. Given the social and legal attitudes against the drug, perhaps some users misrepresent themselves as nonusers. If marijuana genuinely caused a deficit, these incorrectly categorized smokers could lower the average score of the nonusers, making them seem no different from the users. In contrast, in an effort to earn cash through participation, some nonusers might claim to use, potentially altering the scores in this group. A recent study employed urine screens to combat this problem and found 12% of reported levels of use could not be confirmed (Pope & Yurgelun-Todd, 1996). Thus, 12% of the users may have been nonusers or vice versa. This misclassification could seriously alter research results. Urine screens may serve as the best way to avoid contaminated group membership.

INSENSITIVE TESTS The number and characteristics of the participants are not the only aspects of these studies criticized. The tests employed often are too simple to detect marijuana's negative effects. The drug's effects may only appear on more demanding tests. Imagine a cognitive task that required reciting the alphabet. A long and extensive history of cannabis consumption could not impair such an effortless task. Users and nonusers would perform equally well, suggesting no negative impact from marijuana. Yet even a demented person could probably perform this simple exercise. Complex tasks may have a better chance of revealing any cannabis-induced harm. Researchers can counter this critique by including difficult tasks. These investigators may also argue that subtle differences on extremely complex tests may not translate to any practical implications. Difficult cognitive tasks often require discriminating among similar stimuli and responding as quickly as possible. Except for air traffic controllers and fans of video games, it is unclear how many people actually use these skills in daily life. Nevertheless, these tasks prove most sensitive to the subtle changes in brain functioning that may occur after chronic marijuana consumption.

Thus, accurate interpretations of studies that reveal no deficits must consider insufficient sample sizes, biased sampling, contaminated group membership, and insensitive tests. Any study that reveals no deficits associated with chronic marijuana consumption must address these critiques.

Studies That Found Few Marijuana-Related Deficits

One of the first studies to support no marijuana-induced cognitive deficits examined 30 using and 24 nonusing Jamaican men (Bowman & Pihl, 1973). Testing occurred in many environments, including homes, huts, and public areas. This mobile approach to testing helped ensure that even the most impaired individuals could participate. Chronic users had smoked an average of 20 joints a day for at least 10 years. Thus, the sampling does not appear particularly biased toward unimpaired users. The control group included nonusers who were religiously opposed to marijuana. Informants confirmed their abstinence, but no urine screens or hair samples provided additional evidence. Thus, some controls may have actually been users, though it seems unlikely. The assessment included many tests of reaction time, disinhibition, learning, perception, and memory. These tests often show impairment in chronic alcoholics; some have also proven sensitive to marijuana's acute effects. Not one of the 15 measures revealed marijuana-related deficits. The investigators concluded that chronic marijuana consumption had little impact on cognitive abilities.

A comparable study performed in Costa Rica also found no differences between 41 users and 41 nonusers. Participants had an extensive history with the drug, smoking an average of 9 joints a day for 17 years. No biochemical analyses confirmed their status. Users may have claimed to belong to the nonuser group, but such deception may prove unlikely in Costa Rica. The social sanctions against marijuana are not extreme; they were particularly minor in the 1970s, when this study began. For example, certain bistros in the area openly permitted using the drug (Satz, Fletcher, & Sutker, 1976). Participants completed a large assessment battery, including the entire Wechsler Adult Intelligence Scale (an IQ test), numerous tests of memory and psychomotor speed, and a learning task. Many of these tasks have revealed impairments in alcoholics. No significant differences appeared between cannabis users and nonusers. If marijuana created a cognitive deficit, these tests were insensitive to it.

Several other studies also failed to find cannabis-linked cognitive changes, but each had potential weaknesses. One project tested 10 members of a church that views marijuana as a sacrament. These users smoked 2 to 4 ounces of a mixture of marijuana and tobacco each day for an average of 7.4 years. They showed no below-average scores on over a dozen tests of intelligence and neuropsychological functioning (Schaeffer, et al., 1981). The small sample size may preclude finding any meaningful differences. Given the absence of a control group and measures of functioning prior to cannabis consumption, the study cannot reveal if these people may have scored higher if they had not used the drug. Another study revealed no differences between 10 users and 10 nonusers on over two dozen tests related to intelligence and motor performance (Carlin & Trupin, 1977). Use was extensive; participants had smoked daily for an average of 5 years. Nevertheless, the tests were not particularly sensitive and the sample size was too small to ensure that the absence of any effect could generalize to other cannabis smokers.

Additional work also found no marijuana-related deficits, but had potentially biased samples and users with less-extensive drug histories. In a study of Dartmouth seniors, 14 marijuana users, 14 LSD users, and 14 controls completed over two dozen neuropsychological and intelligence tests. Marijuana users did not differ from controls. Incidentally, LSD users differed only on one neuropsychological test (Culver & King, 1974). The absence of an effect may mean little given the purportedly high level of cognitive functioning at this institution. Anyone who showed impairment from the drug might not have remained enrolled in school or may not have been admitted in the first place. In addition, the marijuana group smoked less than twice a week. The tests were not particularly challenging, either. Thus, the absence of differences may stem from a number of reasons.

A different study of 29 using and 29 nonusing medical students, which revealed no differences on 7 of 8 measures of memory and motor skill, suffered from comparable critiques (Grant, Rochford, Fleming, & Stunkard, 1973). Medical students are a relatively intelligent, motivated, compulsive group. People who suffer from marijuana-induced troubles likely could not gain admission to medical school. The users smoked only 3 times per month for 4 years, too. Another study of 26 using and 25 nonusing medical students, which revealed no differences on six neuropsychological measures, had some of the same limitations (Rochford, Grant, & LaVigne, 1977). Users had smoked at least 50 times over an average of 3.7 years, suggesting that many may have smoked less than twice a month. Thus, studies of college and medical students reveal no marijuana-related deficits, but the samples may not represent all users. They may have better cognitive abilities in general, as well as less extensive histories of cannabis use. Researchers may inadvertently exclude impaired users by sampling from students.

One of the largest and most recent studies of marijuana's impact on cognitive functioning looked at changes in mental functioning over 11 years in approximately 1,300 residents of Baltimore (Lyketsos, Garrett, Liang, & Anthony, 1999). A sample this large is certainly beyond critique. Participants were drawn from an enormous epidemiological study of cognitive decline over time. Thus, the sample was not biased to include an inordinate number of well educated or particularly impaired individuals. The classification of marijuana use was based entirely on self-report. Many users may have claimed to abstain, potentially minimizing any group differences. The most important critique of this study concerns the test employed. Participants completed the Mini-Mental State Exam, which, as the authors assert, may be too simple to detect subtle impairments. This brief screening measure detects only the most severe impairments. Easier items include "What is today's date?" and "Where are you?" Even people with Alzheimer's disease and dementia can answer some of these questions correctly.

A Few Differences Deemed Anomalous

Some researchers found a few small differences but concluded that they were spurious for methodological reasons. A study of 60 Jamaican men (30 smokers and 30 controls) used 47 measures of IQ, memory, and motor speed; only 4 showed significant results. Two or three could be expected by chance, as explained below in the discussion of multiple tests performed in studies that reveal differences. Oddly enough, one of these tests suggested better memory functioning in the smokers. Users smoked daily for an average of 17.5 years. The study required a brief stay in the hospital, which may have deterred some of the most impaired users, but the investigators suggested that the free physical exam that accompanied participation may have actually encouraged some users. The tests employed may not have been particularly sensitive, but they did reveal deficits in alcoholics in other studies. These researchers concluded that the evidence for cannabis-induced cognitive impairment was poor (Rubin & Comitas, 1975).

A later study performed in India compared 30 users and 50 nonusers and found only 1 statistically significant memory impairment on 15 different tests. Participants used cannabis an average of 11 times per month for over 5 years, suggesting that use was appropriately extensive. Nevertheless, the tests may not have been difficult enough to reveal marijuana-related changes (Ray, Prabhu, Mohan, Nath, & Neki, 1979).

In summary, quite a bit of research reveals no gross cognitive impairments related to chronic consumption of marijuana. Nevertheless, these studies may have biased, small samples, users with less-extensive drug histories, users who claim to be nonusers in the control group, or tests that require less skill for sufficient performance. Despite these drawbacks, the idea that chronic use of marijuana does not create outrageous neuropsychological problems does receive some support. Other studies that use more complex tasks do suggest marijuana may create cognitive problems, but they suffer from a different set of methodological problems.

Overview of Critiques of Studies That Reveal Deficits in Chronic Users

A review of potential limitations can help the interpretation of studies that reveal deficits in chronic cannabis users. Any deficits found in marijuana smokers may serve as evidence that the drug impairs function, but other explanations remain tenable. Similar to the critiques of studies that find no differences between groups, critics of the research that reveals deficits generally focus on the samples and tests employed. For example, the chronic users may have differed from the nonusers prior to ever smoking marijuana. In addition, the chronic users may have used drugs other than marijuana that might create these impairments. Participants also may have been intoxicated during testing, transforming the study of chronic effects into a study of acute effects. In addition, the multitude of tests employed in many studies may have created some differences simply due to chance. Finally, some effects may reach statistical significance but remain too small to be particularly meaningful. Each of these critiques appears in detail below.

Critiques

DIFFERENCES PRIOR TO MARIJUANA USE Critics of studies that reveal disparities between chronic users and nonusers emphasize a key point about participants. People who choose to smoke marijuana daily for years may be different from those who do not. These differences may have been present long before they started using the drug. Under these circumstances, lower scores found in smokers might not stem from marijuana use. The differences may arise because the groups were unequal for some reason that preceded marijuana use. The best technique for combating this criticism would require randomly assigning people to one of two groups. One group would smoke marijuana every day for years; the other would abstain. This approach has some obvious ethical and practical problems. An alternative strategy requires finding chronic users and nonusers whose cognitive abilities had been assessed before any of them started smoking marijuana. One study has successfully adopted this approach by matching users and nonusers on tests of cognitive functioning that they took in the fourth grade. This study established deficits associated with heavy, chronic use that do not appear to stem from differences that existed before participants started smoking (Block & Ghoneim, 1993). Many other studies lack this careful control. Future research might provide thorough assessments on a large number of young children and then assess cognitive functioning years later, after a small proportion of them become chronic users.

POLYDRUG CONSUMPTION Differences between chronic users of cannabis and nonusers may not arise from marijuana itself. Instead, the users may have consumed other drugs that created impairments. Those who smoke marijuana often use more of other drugs (Earleywine & Newcomb, 1997). Perhaps any identified problems stem from these other substances and not from marijuana. For example, a frequently cited study that reveals deficits in hashish consumers reported more alcohol and opium consumption in the users than in the nonusers (Soueif, 1976). The alcohol and opium, rather than the marijuana, may have created cognitive changes (Fletcher & Satz, 1977). Efficient, ethical ways to combat this critique prove difficult to identify. Extensive, anonymous self-reports may help separate those who use cannabis from those who use other drugs (LaBrie & Earleywine, 2000). Urine screens or hair samples may also help investigators select participants, who use only marijuana or no drugs at all. Focusing on people who use only cannabis may help establish its role in cognitive functioning.

Animal research has had better luck minimizing the possible confounding effect of other drugs. The low rate of substance misuse among primates and rodents permits conclusions about cannabis that remain uncontaminated by polydrug use. Unfortunately, these studies have other methodological problems, including relatively brief exposures to the drug (a year or less) and small sample sizes. Despite these drawbacks, they still reveal a few deficits associated with chronic use. One study found decreased maze learning in rats after 3 months of exposure, but the ability returned after a month of abstinence, (Nakamura da Silva, Concilio, Wilkinson, & Masur, 1991). Other tests of chronic exposure in rats found learning deficits that did not improve, even after months of abstinence (e.g., Stiglick & Kalant, 1982a, b). These animal studies offer compelling evidence that marijuana, and not some other drug, creates problems in learning and memory.

INTOXICATION DURING TESTING Despite researchers' requests and financial incentives, chronic daily users may remain unwilling to abstain from cannabis on the day of an experiment. People who smoke marijuana every day for years may be unlikely to stop simply because a scientist wants them to draw lines, arrange blocks, and memorize words. Without this abstinence, the comparison between chronic users and nonusers essentially becomes a study of intoxication. Given the data on acute effects, studies of chronic effects must clearly occur when all participants are no longer under the influence of the drug. Otherwise, any deficits in chronic users could arise from their current intoxication instead of any impact of long-term consumption. Ensuring this abstinence is extremely difficult when studying people who smoke daily. Inpatient stays in a hospital may help solve this problem. Participants would have to agree to remain in the hospital and avoid drugs for a specified period prior to testing. At least one recent study used this strategy and still found deficits associated with chronic smoking (Pope & Yurgelun-Todd, 1996).

MULTITUDE OF TESTS PERFORMED Another issue in the interpretation of these studies concerns the number of tasks given to the participants and the number of statistical tests that investigators perform. The more tasks employed and the more tests performed, the higher the probability that differences will appear by chance. Every statistical test has a possibility of error. If users and nonusers complete 100 different tests, they may differ on some simply by accident. The more tests, the greater the likelihood of such accidents. Initial studies of cannabis's cognitive effects cast a wide net in search of impaired functioning. This approach, however, increased the probability of finding differences by chance. For example, a classic study that showed marijuana-induced harm performed almost 100 statistical tests (Soueif, 1976).

Most research journals report differences as statistically significant only if they are big enough to be unlikely to have occurred by chance. If nonusers perform 50% better than users on a test, the odds of the difference occurring by chance may be small. But if the same number of participants differs by only 20%, 10%, or 1%, the probability of the difference arising by accident increases. Convention dictates that statistically significant differences must be large enough to only occur by chance less than 5 times out of 100. (The ubiquitous "$p < .05$" appears in research when findings satisfy this convention.) Nevertheless, 5 times out of 100, or 1 in 20 of such findings may be anomalies. Hypotheses focused on a few key cognitive functions can help minimize this problem by minimizing the number of tests performed and thus the potential number of chance findings.

SIZE AND MEANING OF STATISTICALLY SIGNIFICANT DIFFERENCES The last critique of these studies concerns the magnitude and meaning of statistically significant effects. With big enough samples, very small differences in test performance can be statistically significant. The meaning of such small effects remains debated. For example, a frequently cited study showed that urban Egyptian users had worse memory for numbers than nonusers. Although the groups did differ statistically, the nonusers recalled, on average, 2.94 digits. The users remembered 2.75 digits (Soueif, 1976). This average difference is less than a quarter of a word. Critics find these small effects meaningless (Zimmer & Morgan, 1997). Studies that reveal marijuana-related deficits can prove most compelling if the impairments are large and involve important aspects of thought and memory.

Studies Showing Marijuana-Related Differences

Accurate interpretations of studies that reveal deficits must consider differences prior to use, polydrug consumption, intoxication during testing, multiple tests employed, and reporting of small or inconsequential differences. Any study that reveals deficits associated with chronic marijuana consumption must address these critiques. Generally, these studies suggest that long-term use of cannabis does not lead to overt signs of gross intellectual impairment. Nevertheless, subtle problems on difficult tasks do arise.

One of the first and largest sets of studies assessed over 1,600 Egyptian prisoners (Soueif, 1976). Ten of 16 measures of cognitive abilities showed differences, but two (time and distance estimation) revealed better performance in the smokers. The most differences appeared for the educated people. Users and nonusers with at least a high school education differed on more tests than those with less schooling. The critiques listed above apply to this series of studies. They did not report any measures of cognitive functioning prior to use, failing to rule out potential differences between users and nonusers before they consumed cannabis. Many of the smokers also used opium, which may have contributed to their lower scores (Fletcher & Satz, 1977). Nothing prevented the users from smoking immediately prior to testing, but being imprisoned may have minimized this confound. The tests were not particularly numerous, but some of the statistically significant differences remained small. For example, as mentioned, users and nonusers differed significantly at recalling numbers, but by less than one digit (Soueif, 1976).

Another series of studies performed in India also revealed cannabis-related deficits. A comparison of 23 users and 11 controls revealed poorer functioning on measures of IQ, motor speed, and time perception. Participants completed 8 cognitive tests, 6 of which showed differences, suggesting that the results were not likely chance anomalies. Also, the differences were large and potentially meaningful (Cohen, 1990), particularly on the measures of IQ (Wig & Varma, 1977). Yet some of the standard critiques still apply. Cognitive differences prior to use were unknown. Consumption of other drugs was not reported. Only some of the participants were hospitalized to ensure sobriety during testing. All these results may stem from differences in cognitive functioning that existed before the participants ever began using the drug. Differences also may arise from current, acute intoxication in some people.

Another study performed in India on 25 smokers, 25 bhang drinkers, and 25 controls revealed significant differences on time perception, memory, size and time estimation, motor speed, and reaction time. (Bhang is a beverage made from marijuana, milk, sugar, and spices.) The investigators did not report cognitive abilities prior to use. Participants differed little in alcohol consumption. The study did not report assessments of other drug use. The researchers claim that participants did not consume cannabis for 12 hours prior to testing, but the techniques employed to ensure this abstinence remain unclear.

Participants in this study completed only 10 tests, 8 of which revealed differences. Some deficits remained small. For example, a memory test that required recognizing pictures differed by less than one picture. Nevertheless, other effects were quite large. Reaction time for ganja smokers was twice as long as for controls (Menhiratta, Wig, & Verma, 1978). The measure of reaction time, however, was quite unusual. Most reaction time measures require pressing a button as quickly as possible after seeing a light or hearing a tone. The measure in this study asked participants to report the first idea that came to mind when they heard a word. The meaning of a slow response on this task is unclear. Smokers could have responded more slowly for many reasons unrelated to cognitive abilities. The failure to control for previous cognitive abilities, polydrug use, and intoxication warrants cautious interpretation.

An intriguing 10-year follow-up of some of these participants (19 smokers, 11 bhang drinkers, and 15 controls) replicated the memory, motor, and reaction time findings. Although previous cognitive abilities and polydrug use remained uncontrolled, a hospital stay helped ensure sobriety during testing this time. The same number of tests were employed. The authors failed to present the data in a way that permits the computation of the exact size of the effects, but some appear potentially meaningful. For example, the controls did twice as well as the bhang drinkers on remembering digits in reverse order. The same curious reaction time measure also showed great differences between users and nonusers. Nevertheless, we do not know if the groups differed in abilities before they started using cannabis and have no evidence that the use of other drugs did not create the deficits.

Another long-term follow-up study found differences in a sample that had previously not shown any marijuana-induced changes in cognitive functioning. The investigators tracked down the participants from the Costa Rican study that initially revealed no differences (Satz, et al., 1976; see the section in this chapter entitled "Studies That Found Few Marijuana-Related Deficits"). They then administered the same battery and some new measures. These investigators found differences on 3 new tests. Users and nonusers showed comparable scores on the same tests they had completed 12 years earlier. Their scores had deteriorated little in the 12 years. The three new tests, however, showed superior performance in nonusers. One test tapped retrieval from memory; the other two required sustained attention and concentration. We do not know if the groups differed in these abilities prior to using cannabis. They did not differ in alcohol or tobacco use, but we do not know about other drugs. Investigators asked participants to abstain from alcohol and marijuana the day before the test, but adherence to these recommendations was not assessed. The tests were numerous, and differences were small (Page, Fletcher, & True, 1988).

An additional follow-up began 5 years later. The data revealed group difference only in older participants. These people were about 45 years old and had smoked cannabis for an average of 34 years. They completed 4 tests of memory and 8 of attention. Two memory tests and two attention tests revealed differences. Yet, differences prior to use remain unknown. Those who chose to smoke regularly may have had attentional or memory deficits prior to ever using the drug. Urine screens confirmed user status as well as an absence of use of other drugs and no intoxication during testing. The effects were in the medium to large range. These data suggest that long-term use creates specific memory and attention problems, but differences prior to ever using the drug cannot be completely ruled out.

Although many studies of college students have shown no marijuana-related differences (Culver & King, 1974; Grant et al., 1973; Rochford et al., 1977), two that used different tasks revealed some memory problems. A study of 25 nonusing undergraduates and 25 who smoked at least twice a week suggested that marijuana interferes with transferring information into long-term memory. We do not know their level of ability before they started using cannabis or their use of other drugs. Participants reported that they were not high, but abstinence was not confirmed. There were no other tests reported, and the effects qualified as large (Gianutsos & Litwack, 1976). Another study of 26 college students who had used daily for at least 6 months and 37 who had never smoked marijuana revealed large deficits in memory and learning. These participants did not differ on drinking habits, but use of other drugs was not assessed. The investigators repeated the experiment with 21 users and 18 controls and found comparable marijuana-related problems in memory. This time, those who used other drugs were not allowed to participate (Entin & Goldzung, 1973). Again we have no knowledge of cognitive ability prior to the initiation of use, and intoxication during testing was not controlled. The effects were in the medium to large range.

Superior Performance in Users

A couple of other studies revealed the unexpected superior performance of users over nonusers. Results like these have never inspired anyone to recommend cannabis as a way to enhance cognitive abilities, but they do cast doubt on the idea that chronic marijuana use is detrimental. One of the first reports of superior performance came from an examination of 11 nonusers and 11 people who smoked 3 to 5 times per week for an average of 4 years. The groups did not differ on over a dozen cognitive tests. The users performed better on one test of general cognitive functioning, but the large number of analyses suggest cautious interpretation. The same researchers found superior performance by users on 8 of 11 cognitive tests in another sample (Weckowicz, Collier, & Spreng, 1977).

The superior performance of marijuana users may stem, in part, from the nature of the measures. Unlike the tests employed in most studies, these focused on originality and novel thinking rather than speedy processing of information. These data do not, however, mean marijuana smokers are more creative than others. The groups may have differed in originality prior to use. Polydrug use was rampant among the smokers, raising the possibility that another drug may have created these effects. The users may have been high during testing. The researchers used many tests, but some effects were in the medium to large range. The biggest critique, however, involves the recruitment of participants. This study did not involve a random sample of users and nonusers. Instead, users supplied the names of other users to participate. This process may have biased results. If one creative user recommended a bunch of creative friends, the results could easily suggest chronic use led to superior performance erroneously.

Successful Control for Level of Functioning Prior to Use

In one of the few studies to control for cognitive abilities before onset of drug use, investigators compared 144 users and 72 nonusers matched on the Iowa Test of Basic Skills, which they had taken in fourth grade. The investigators compared heavy (7 times per week), intermediate (5 or 6 times per week), and light (1 to 4 times per week) users to the nonusers. The heavy users had smoked for an average of 6.2 years. Participants completed 17 measures of memory, learning, and motor skills. Nonusers performed better than heavy users on 4 tests: one of quantitative skill, one of verbal expression, and two measures of memory for words that are easy to imagine.

The light and intermediate users were indistinguishable from nonusers on all but one test. Unexpectedly, intermediate users performed better than nonusers on a test of concept formation. (This test requires looking at pictures of two families. Pictures of new people then appear and participants must guess which family they belong to.) There were no other significant differences among groups. Heavy users performed as well as nonusers on 13 tests (Block & Ghoneim, 1993). Despite admirable control for differences prior to use, consumption of other drugs differed dramatically among groups. Investigators requested abstinence during testing but did not ensure it. Participants did perform many tests, but all effects were large (including the superior performance of the intermediate group on concept formation). Unfortunately, the effects may have stemmed from intoxication. This study offers potential support for marijuana-induced impairment and also may imply that intermediate and light use does not create cognitive deficits.

Comparing Light and Heavy Users

One group of investigators reasoned that those who try marijuana may differ from those who do not on a number of variables. These inherent differences may contaminate results. Therefore, this research team compared 65 chronic users and 64 light users in the search for marijuana-related cognitive problems. Chronic users smoked at least 22 days in the previous 30; light users smoked 9 or fewer days. Chronic users performed worse on 2 of 7 measures, including a neuropsychological test of disinhibition (the Wisconsin Card Sorting Test) and a test of learning, recognition, and recall (the California Verbal Learning Test). These effects were in the small to medium range. For example, on one measure of verbal recall, the groups differed by less than a word.

A few other effects appeared when the investigators examined men and women separately. Among the men, heavy and light users differed on some subscales of tests related to spatial memory. Heavier-using men performed worse on the recall of pictures. The groups also differed on the Stroop, the color-naming test of disinhibition. Another effect appeared when the investigators divided participants into groups based on IQ. Heavy and light users with low scores on verbal intelligence differed on verbal fluency (Pope & Yurgelun-Todd, 1996).

The usual critiques apply to this study. We do not know about differences between these groups prior to their use of marijuana. Differences in the use of other drugs, however, was minimized based on self-reports and the urine screens. People who used other drugs were systematically excluded. The possibility of intoxication was virtually eliminated with a hospital stay that included observation to ensure no one smoked marijuana prior to testing. Despite these methodological strengths, the number of statistical tests performed became quite high once groups were divided by gender and IQ and tasks were divided into subscales. Over 60 statistical tests appear, increasing the possibility of chance findings. In addition, effects only ranged from small to medium.

Adolescent Samples

All the studies listed so far were performed on adults. The potential impact of cannabis in adolescence has not received appropriate attention.

Given the development of many cognitive abilities during this stage, chemical insults could prove quite severe. Only a couple of studies of adolescents appear in the literature (Schwartz, 1991; Schwartz, Gruenewald, Klitzner, & Fedio, 1989). They are frequently cited as evidence for cannabis-induced impairment but have drawn criticism for methodological flaws (Zimmer & Morgan, 1997). These data are particularly important because so little is known about marijuana's impact on young people.

In other work, memory troubles appeared on 2 out of 7 tests in 10 cannabis-dependent adolescents in a drug treatment program when compared to 17 controls. Some but not all of the usual critiques apply to this study. We do not know if these adolescents would have differed on these tasks before they started using. Some participants had used phencyclidine, which causes cognitive troubles (Cosgrove & Newell, 1991). They probably were not intoxicated during testing; the assessments occurred in a treatment center. Only 10 tests were reported, and the effects on both of the tasks that revealed differences were large. Nevertheless, over half of the control group was not attending the treatment program. We do not know the impact of participating in this treatment program, but the investigators mention that the extreme emotional reactions associated with admission may interfere with cognitive abilities (Schwartz et al., 1989). An improved study of adolescents would include more appropriate controls. Despite the many limitations of this study, the National Organization for the Reform of Marijuana Laws recommends no cannabis use in children and adolescents (NORML, 1996a). Further research on adolescents would fill an important gap in this literature.

Event-Related Potentials

A body of evidence reveals that chronic, heavy marijuana smokers may not perform as well as nonsmokers on complex cognitive tasks. Another technique for assessing changes in information processing involves changes in an electroencephalogram or EEG. One change associated with chronic consumption of marijuana concerns deviant brain waves that appear during a difficult task (Solowij, 1998). Brain waves that occur in response to a particular event (like a light or a tone) are called event-related potentials. These alterations in brain waves can reveal information about cognitive processes. A series of studies looked at these brain waves and found deviations in chronic, heavy users of marijuana that suggest they do not process information as accurately or rapidly as nonusers.

The task employed in these studies of event-related potentials was very difficult. People had to listen to a series of tones. Some sounded in the left ear and some in the right ear. The tones were also long or short and high or low in pitch. Participants had to press a button as fast as possible in response to the longer tones of a specified pitch in the correct ear. For example, the investigators might ask a participant to press the button only in response to high-pitched, longer tones presented in the right ear. People can easily discriminate between tones in their left and right ears, but the pitch and duration were more challenging. The high and low tones were fairly similar. They corresponded to the notes C and E on the same scale. The lengths were also comparable. The long tone was only 51 milliseconds (.051seconds) longer than the short tone.

In a series of studies using this task, chronic cannabis users showed problems discriminating between the tones. They failed to press the button after the target tone more frequently than nonusers. They also pressed the button in response to nontarget tones more often than the nonusers. In addition to making more errors, the chronic users showed deviant event-related potentials. Nonusers tended to show large event-related potentials after the target tones, but their brain waves did not change after the irrelevant tones. The large changes suggest cognitive processing of the relevant tones. The absence of changes suggests that participants quickly identified the irrelevant tones and did not process them. Compared to the nonusers, chronic users showed slower changes in response to the target tones and larger changes after the irrelevant tones. The brain wave changes in chronic users suggest that they had a harder time separating the relevant from the irrelevant tones. Thus, chronic users appeared to have more trouble distinguishing between these different stimuli.

Most of the usual confounds for research that reveals differences between users and nonusers were controlled in this study. They may have differed in abilities prior to ever using cannabis, but they performed equally well on a reading test that correlates highly with general IQ. (Many investigators assume that IQ is very stable, suggesting that the groups likely did not differ before they started using cannabis.) Differences in the use of other drugs were minimized by eliminating any potential participant who reported using any other drug more than once a month. The users and nonusers also had comparable drinking habits, educations, and ages. Urine screens confirmed that participants were not high while performing the tests.

Further work has added support to the hypothesis that chronic marijuana use served as the genuine source of the poor discrimination performance and deviant brain waves. A subsequent study showed that at least one type of deviant event-related potential grew worse with longer durations of chronic cannabis use (Solowij, 1998). This potential (processing negativity or PN) was a response to irrelevant tones that was larger in chronic users than nonusers, confirming problems with separating relevant from irrelevant stimuli. Another type of potential (P300) occurred more slowly in all users, regardless of how long they had smoked cannabis. This result suggests that chronic users may take longer to process some information.

Another subsequent study examined these same variables in people who had used marijuana regularly for at least 5 years but who had quit an average of 2 years before the experiment (Solowij, 1998). These ex-users still performed more slowly than nonusers at pressing the button after the target tone, but they did not hit the button incorrectly any more often than nonusers did. Their brain waves improved, too. Their PN responses to the irrelevant tones were not as large as those found in the chronic users who were still smoking marijuana, but they were still not as small as those found in the nonusers. These results suggest that the ability to separate relevant from irrelevant stimuli may take longer to recover. In contrast, the slowing of the P300 was no longer present in the ex-users. Apparently, this aspect of processing recovers more quickly. These comparisons between users, nonusers, and ex-users offer considerable support for the hypothesis that chronic marijuana use leads to these cognitive changes.

Conclusions

Acute intoxication with marijuana clearly impairs a number of cognitive tasks, including aspects of memory, perception, reading aloud, arithmetic, and complex reaction time. Acute intoxication may or may not alter other facets of thinking, including simple reaction time, disinhibition, and vigilance. Some research shows deficits in these tasks, but other studies do not. More research with larger samples and multiple measures of these abilities could help establish whether or not marijuana impairs reaction time, disinhibition, and vigilance. Other cognitive abilities remain intact during intoxication, including the ability to learn simple tasks and remember material mastered prior to using the drug. Users should avoid certain tasks during intoxication if optimal performance is essential.

Studies of the impact of chronic, heavy use of marijuana are fraught with numerous confounds. Despite the many limitations of the different studies, a few conclusions appear tenable. Long-term exposure to cannabis probably does not affect gross intellectual functioning. Nevertheless, the ability to perform quickly on elaborate tasks likely decreases with chronic use. Studies of event-related potentials reveal that the processing of information differs after years of regular cannabis consumption. These results suggest that chronic users may not provide the best performance on complicated tasks that require speedy responses. These deficits imply some alteration in brain function that accompanies chronic exposure to marijuana. The implications for these effects on the brain and nervous system appear in chapter 7.

5
Subjective Effects

Perceptions change during marijuana intoxication. Time and space appear distorted. The senses seem more sensitive. Higher functions like thought, memory, and spirituality can alter, too. Some of these changes stem from the pharmacological properties of the cannabinoids. Others arise from the expectations of the user, the demands of the environment, or the attitudes of the culture where the drug is ingested. These factors can combine in unpredictable ways to create odd experiences. This chapter describes some of the difficulties associated with assessing subjective experience and addresses marijuana's perceived effects on time, space, and the senses, as well as higher functions like emotion, thought, memory, sexuality, spirituality, and sleep. An overview of these effects is given in the following list:

Perceptions
Time slows
Space appears more vast
Senses appear enhanced

Emotions
Euphoria increases
Relaxation increases
Feelings seem stronger
Fear increases at high doses

Thoughts
Focus on the present increases
Forgetfulness increases

Sexuality
Orgasms appear enhanced
Responsiveness appears enhanced

Spirituality
Openness to experience increases
Sense of the divine increases

Sleep
Improves at low doses
Shows impairment at high doses

Undesirables
Concentration appears impaired
Depersonalization
Eyes redden
Mouth and other mucous membranes lose moisture

Most people cannot find the words to explain their sensations. Describing simple changes in perception or emotion remains difficult enough during the most sober and wakeful moments. Add the confounding effects of drugs, and words can fail to portray experience. Despite the difficulties associated with describing consciousness, regular users of cannabis often develop their own jargon to depict marijuana's effects. Novel terms describe the intoxication that different strains of marijuana may produce. Some varieties purportedly create "heady" or cerebral experiences. Others lead to "mellow" or sedating effects. Another type may earn the label "laughing grass" because its users find everything funny. These fine distinctions among subjective states parallel the subtle discriminations that wine connoisseurs make when judging the latest vintage. In fact, cannabis competitions in Amsterdam often rank products based on these subjective effects.

Some of the first attempts to describe the phenomenology of marijuana intoxication appeared in literature rather than science. Gautier, Ludlow, Baudelaire, Ginsberg, and many other authors have written thousands of words in an effort to depict cannabis's effects. Scientific researchers have reported case studies of intoxication to try to describe the drug's impact, too (e.g., Moreau, 1845). A few recurrent themes appear in these works. First, reactions to cannabis vary dramatically from person to person. Some people truly detest the experience, reporting painful self-consciousness, disorientation, and paranoia. They rarely use the drug more than once. Research may underestimate the frequency and severity of these aversive reactions, because most studies focus on experienced users who enjoy the drug. These individuals often appreciate cannabis immensely. Some of their reactions sound similar to descriptions of religious ecstasy. This huge variation in responses suggests that no one person's intoxication experience is typical.

Second to the vast individual variation in reactions, all reports emphasize that attitudes and surroundings contribute to the subjective experience of the drug. Individuals who expect to have a pleasant time and arrange for cozy, safe, relaxed surroundings frequently report positive effects from the drug. In contrast, those who anticipate a fearful experience and use the drug in an uncomfortable setting can report panic and suspiciousness. Baudelaire (1861) described the importance of these circumstances over 140 years ago, recommending that users of hashish find a nice spot outdoors or a well-decorated room with a little music. Zinberg (1984) refers to the relevant environment as the setting and the individual's expectations as the set. He asserts that both set and setting can contribute to the impact of any psychoactive drug. Many studies support his ideas.

Because environment and expectations contribute to the subjective experience of marijuana, the study of the drug's effects requires diverse methods and careful interpretations. No single experiment can depict the phenomenology of intoxication perfectly. Detailed reports of individual experiences, including the literary work of the authors above, can help generate hypotheses about the drug. Nevertheless, the impact on a few authors may not generalize to everyone. This literary work inspired the first formal research on larger samples of people. These studies focused on self-reported marijuana effects (Halikas, Goodwin, & Guze, 1971; Tart, 1971). Instead of examining a few long, literary narratives about intoxication, these researchers employed larger groups of users who answered structured questions about their experience. Their results appear in detail below. Although these data reveal a great deal about the drug, they can only tell part of the story of cannabis intoxication. People's memories about their moods and emotions may not accurately reflect their sensations, so some effects may only appear reliably when users consume the drug in the laboratory.

Memory problems may not be the only source of error in reports of marijuana's impact. People can mistakenly think that the drug has changed their mood when other activities may have altered their feelings. While smoking cannabis, people often behave in ways that would change their emotions, even in the absence of the drug. Listening to music, savoring favorite foods, and enjoying other activities stereotypically paired with marijuana consumption may provide some of the euphoria usually attributed to the substance. People may say that marijuana makes them happy, when, in fact, it is the chance to watch TV, have sex, or walk in the woods that actually improves their moods. In addition, some cannabis effects may stem from expectations rather than pharmacology. People may expect a drug to make them happy and may end up feeling happy simply as a result of the expectation. These potential limitations of research on self-reports inspired many laboratory studies of cannabis, hashish, and isolated THC.

Laboratory administrations of cannabis or cannabinoids can validate self-reports and case studies. People may describe experiences in the laboratory that parallel accounts of intoxication in the field. These experiments can also help disentangle pharmacological effects from expectancies. Crafty researchers may administer credible placebos, including marijuana with no THC, to see if subjective sensations alter in response to the mere idea of smoking cannabis. Any changes in reaction to the placebo obviously must stem from beliefs about the drug rather than any chemical. In fact, at least part of the subjective effects of the drug arise from these expectations. This research approach can help isolate THC's impact in a way that self-reports about intoxication outside the laboratory cannot.

Yet these laboratory studies have limitations, too. Laboratory settings are often artificial. Good experimental design usually requires keeping all aspects of the environment the same for those who use the drug and those who receive the placebo. Then any difference in experience must stem from the cannabis itself. Unfortunately, this sort of control may lead to administering the drug in settings that fail to generalize to the world outside the laboratory. For example, research might focus on a lone smoker in a sterile room completing questionnaires while surrounded by sober experimenters. This approach may not provide the best data on emotional responses to the drug. Yet bringing a group of friends into the laboratory to share a pipe and discuss whatever comes to mind leaves too much to chance. The emotional experience may stem from the particular mix of people, the topics they haphazardly choose to discuss, or the drug itself. Thus, each approach to understanding marijuana intoxication has limitations. Only the sum of research using different methods can help depict the experience. Examining self-reports and laboratory work may serve as the only way to understand the phenomenology of marijuana intoxication, short of Ginsberg's (1966) recommendation of ingesting the drug oneself.

One of the first formal investigations of the phenomenology of marijuana intoxication asked more than 200 questions of 150 people who had used the drug at least a dozen times (Tart, 1971). Participants were mostly college students from California. The questionnaire described specific experiences that might occur during intoxication, such as feeling euphoric or forgetting things. Participants rated how frequently they had these experiences after using marijuana. The rating scale extended from "not at all," to "sometimes," "very often," and "usually." Participants also rated how high they had to be in order to first notice the effect. Like any research, this study is a product of its era. The questions employed the jargon of the late 1960s. For example, people were asked how they feel when they "turn on" (an expression for using marijuana). Other items referred to understanding people's "games" (the social scripts that guided their behavior) and "hang ups" (their troubles).

Despite using the slang of a different era, this study revealed a lot about intoxication that should apply today. The research identified over 30 experiences that were characteristic of the marijuana high. Tart (1971) defined a characteristic effect as one that at least half of the people indicated that they experienced very often or usually. He also identified many effects that did not occur as often as the characteristic ones but still seemed quite common. Common effects were those that half of the participants (or more) reported experiencing at least sometimes. These data confirmed many of the reports in early literary works and case studies. Subsequent laboratory research that actually administered THC supports many of Tart's (1971) seminal findings. Other researchers using comparable interviews and questionnaires also added to the understanding of the cannabis intoxication experience. The results of this work fall into a number of different categories based on different effects. Some focus primarily on perception; others involve higher functions like emotions and spirituality.

Perception

Time

Literary accounts of intoxication frequently mention a distorted flow of time (Gautier, 1846; Ginsberg, 1966). Early case studies of cannabis's effects also emphasized the feeling that time slowed. A thirty-second commercial or three-minute rock song may seem to last markedly longer after using marijuana. Tart's (1971) participants reported that the characteristic effects of cannabis included the sense that time passes very slowly. This effect appeared at moderate levels of intoxication or more. The participants also suggested that they gave little thought to the future because they focused primarily on the current moment. A separate sample of 100 people who had used cannabis at least 50 times confirmed this effect on time perception (Halikas, et al., 1971).

Laboratory methods for assessing the subjective experience of time also support subjective slowing. One task, time estimation, requires that individuals wait for a period and then guess the amount of time that has passed. Participants who have smoked marijuana overestimate the duration. After waiting 30 seconds, intoxicated people might report that a minute has passed. In another approach, known as time production, people must press a button once, wait until they think a specified period has passed, and then press the button again. For example, they could hit the space bar of a computer once and then again when they think 30 seconds has passed. This technique also reveals subjective slowing. After smoking marijuana, people produce shorter intervals. They might wait only 20 seconds between presses when asked to wait 30 seconds. They seem to think that more time has passed than actually has. A review of a dozen of these laboratory studies suggests that time clearly slows during intoxication, with larger effects for longer intervals (Chait & Pierri, 1992). Alcohol also appears to have this effect (Lapp, Collins, Zywiak, & Izzo, 1994). No studies address the impact of both marijuana and alcohol on time perception. Perhaps they combine to alter time more dramatically.

Space

Moreau's (1845) early studies of hashish intoxication revealed a deviant sense of spatial relationships. Members of the Hashish Club perceived distances inside the room as markedly larger than their actual size. Tart's

(1971) sample also reported that their experience of distance changed after smoking marijuana. They commonly found that the space between people seemed larger, particularly at moderate levels of intoxication or more. They also felt that the distances they walked were quite different. This phenomenon was a characteristic effect reported by over 75% of the sample at low to moderate levels of intoxication. Laboratory work confirms these deviant perceptions of space. In a study where people drove a specified path after smoking marijuana, they overestimated the distance they traveled. This effect, however, may have been confounded with time distortion. If an intoxicated person feels that a particular drive took more time, he or she might assume it included a longer distance (Bech et al., 1973). Thus, both time and space appear altered after using marijuana.

Vision

Changes in vision often accompany marijuana intoxication. Intoxicated individuals report enhanced visual acuity and depth perception, but laboratory studies suggest impairment in these same abilities. Perhaps marijuana makes people feel that certain perceptions are enhanced, even when they are not. Tart's (1971) participants reported an enhanced ability to identify patterns in meaningless visual stimuli after using cannabis. This characteristic effect occurred when users were strongly intoxicated. Confirming this experience in the laboratory would prove quite difficult. It is unclear how to measure the patterns or their meaning. Full-blown visual hallucinations, where individuals perceive objects that clearly do not exist, are extremely rare even at high doses of cannabis. Nevertheless, 9% of Tart's (1971) sample considered hallucinations a usual effect that began at very high levels of intoxication. Another study of experienced users revealed that a few (4%) claim to see visions (Halikas et al., 1971). Ancient Asian texts suggested that these visions appear at large doses (Abel, 1980). This effect also proves difficult to investigate in the laboratory, but many clinical case studies report hallucinations following cannabis consumption at large doses.

Although hallucinations are infrequent, altered perceptions of existing stimuli appear quite often during intoxication. Gautier (1846) and others from the Hashish Club wrote extensively about distorted visual perceptions of their natural surroundings. For example, an ordinary room suddenly seemed darker and more unusual after eating hashish. These descriptions may suffer from the exaggeration and drama associated with literature from the period, so other research may be more trustworthy. About half of Tart's (1971) participants reported seeing auras or lights around people's faces at extremely high levels of intoxication. This sort of visual perceptual aberration may stem from the popularity of auras in that era, some sort of expectancy, or an actual pharmacological effect of the drug. A sample of more than 200 Canadians confirms that users frequently report visual effects (Adamec, Pihl, & Leiter, 1976). Yet no laboratory work directly addresses these experiences.

Tart's (1971) sample also reported an improved ability to imagine pictures and objects, starting in the low to middle ranges of intoxication. Laboratory studies have failed to confirm these reports. Perhaps the perception of the ability to imagine does not reflect the true ability. For example, participants who used imagery in a learning and memory task were asked to describe the images that they used. Judges rated the intoxicated people's descriptions as less vivid than the other group's (Block & Wittenborn, 1984b). These results may mean that marijuana intoxication makes people think that they are better at imagining objects when in fact they are not. Nevertheless, these data also may mean that intoxicated individuals do not describe images as vividly. The deficit may not be in the ability to imagine, but in the ability to articulate the imagined. Thus, research confirms that people think that cannabis improves their imaginations, but has yet to confirm any actual positive changes.

The perception of colors also alters after smoking cannabis. Tart's (1971) sample reported that they commonly saw new colors or more subtle shades of color during intoxication. Participants claimed this effect occurred when they were at least fairly high. Laboratory research reveals the opposite effect, contradicting these self-reports. After smoking marijuana, participants did a significantly poorer job of distinguishing between different hues (Adams, Brown, Haegerstrom-Portnoy, & Flom, 1976). The poorest discriminations appeared in the blue range of the spectrum. Thus, the perceived effects of marijuana on color perception do not appear to parallel the actual effects. Perhaps the drug merely makes users think that this skill has increased.

Cannabis intoxication may also alter perceptions of depth. Tart's (1971) participants reported that pictures and images took on an added three-dimensional appearance after smoking marijuana. This effect began in the low to middle range of intoxication. A fairly recent and intriguing case study supports this effect (Mikulas, 1996). The author describes a

42-year-old physician who had always perceived the world as flat and two dimensional. While smoking cannabis and viewing mountains, he suddenly perceived depth. The effect dissipated but then returned as his eyes and mind were eventually trained to see in three dimensions. Laboratory studies have yet to confirm this effect on depth perception. Oddly enough, intoxication in the laboratory actually decreases the illusion of three dimensions created by certain pictures presented with a stereoscope (Emrich et al., 1991).

Finally, another alteration in vision concerns marijuana's impact on intraocular pressure, the force within the eye. Robert Randall suffers from glaucoma, a disorder associated with increased intraocular pressure. He sees halos around lights when the pressure in his eyes climbs. Marijuana reduces this pressure and makes the halos disappear (Randall & O'Leary, 1998). This improvement in vision may serve as one of the few examples of cannabis normalizing a perceptual process. Laboratory research confirms the decrease in intraocular pressure, but no studies have actually assessed the disappearance in the perception of halos around lights. Large studies of subjective reports have yet to ask about this effect. Nevertheless, data reveal that people think marijuana can enhance some visual processes, and laboratory research suggests it actually impairs some of them.

Hearing

Marijuana intoxication may alter the perception of sounds. Balzac (1900) mentions hearing bells after eating hashish. Moreau (1845) found people grew more sensitive to noises and music during intoxication. Gautier (1846) claimed to hear celestial chords at the Hashish Club. Tart's (1971) participants reported a few characteristic effects related to hearing. These included understanding the words to songs better, detecting more subtle changes in sound, and perceiving greater separation between sources of sound. Users claimed that cannabis improved their understanding of lyrics, even at relatively low levels of intoxication. No laboratory studies confirm these results. Given the lyrics of many contemporary songs, any marijuana-induced improvement in comprehension might seriously harm sales.

Improved perception of subtle changes in sound includes the idea that notes of music sound more distinct or that rhythms seem more clear. Almost all of Tart's (1971) sample (95%) reported this enhanced auditory acuity very often or usually. These effects began at low levels of intoxication, too. Separate reports from 100 American users and 200 Canadian users confirmed that people believe their hearing improves after using marijuana (Adamec et al., 1976; Halikas et al., 1971). Laboratory research on the auditory acuity and the perception of differences in tones is quite advanced, but these procedures have never been applied during cannabis intoxication. Thus, validation of marijuana's purported improvement of hearing awaits further research.

A common, related auditory effect concerns sound taking on visual, colorful qualities. Researchers call this confusion of one sense for another "synaesthesia." The idea of music having color obviously mixes the visual and auditory domains. This effect appeared only at high levels of intoxication in Tart's (1971) sample. More than 200 Canadian users confirmed this synesthesia (Adamec et al., 1976). Laboratory studies have never assessed this phenomenon.

The last auditory improvement reported by Tart's (1971) participants concerns greater spatial separation between sound sources. Users suggested that when they listened to music, they felt that the instruments sounded further from each other. This phenomenon began at moderate levels of intoxication. This effect likely gave the impression of improved stereo qualities, with different sounds emanating from different locations in the room. This effect may fit the recurring theme of enhanced sensory experiences from marijuana. Researchers have developed paradigms for testing the perceived distance between sound sources, but none have been applied during cannabis intoxication. In general, appreciation for sounds appears to increase with intoxication, but researchers have yet to confirm or disprove the effect in the laboratory.

Touch

Altered perceptions of tactile stimuli serve as a hallmark sign of marijuana intoxication. Literary accounts of intoxication consistently emphasize an altered sense of touch. Baudelaire (1861) mentions this effect in "The Poem of Hashish." The majority of Tart's (1971) sample (65%) reported that they usually experienced a more exciting, more sensual sense of touch after using cannabis. This effect appeared at the lower middle ranges of intoxication or more. In addition, they reported that touch took on new qualities. Over half (55%) of the sample experienced novel tactile sensations when intoxicated. This finding fits the general theme of perceptions of enhanced sensations. A separate sample of 100 experienced users confirmed that marijuana enhanced their sense of touch (Halikas, Weller, & Morse, 1982); more than 200 Canadian users also reported this effect (Adamec et al., 1976). These tactile sensations may contribute to marijuana's legendary enrichment of sexual experiences, which appears in more detail in the discussion of higher functions.

Laboratory studies on the thrill of touch after using cannabis have not appeared, though conducting them should not prove difficult. Blindfolded subjects could smoke marijuana or a placebo, feel various tactile stimuli, and rate the enjoyment associated with each. Higher ratings in the marijuana group would support an enhanced sense of touch associated with intoxication. Despite self-reports and literary examples of enhanced tactile senses, experiments suggest that related abilities may actually suffer during intoxication. One study asked participants to solve a puzzle using only their sense of touch. Participants had to place ten wooden shapes (a square, a cross, a diamond, etc.) into appropriate slots on a puzzle board. They took significantly longer to solve the puzzle after smoking cannabis. Participants who smoked placebo showed no deficit (Maccannell, Milstein, Karr, & Clark, 1977). These results do not mean that tactile perception is necessarily impaired. Perhaps the enhanced novelty and excitement of touch interferes with the ability to recognize shapes from only tactile information. Perhaps intoxicated individuals have less motivation to work quickly on such a task. Nevertheless, marijuana's purported facilitation of the sense of touch clearly does not help people solve tactile puzzles.

Taste

Gautier (1846) claimed that the simplest water tasted like exquisite wine after eating hashish. Marijuana's legendary impact on appetite has generated many humorous depictions of the "munchies." Tart's (1971) sample reported that the drug made taste sensations take on new qualities. This effect began even at low levels of intoxication. Other samples of experienced users also reported enhanced appreciation of tastes (Adamec et al., 1976; Halikas et al., 1982). Laboratory studies fail to reveal improvements in the ability to taste classic sour, sweet, salty, or bitter substances (Mattes, Shaw, & Engelman, 1994). Thus, intoxicated people may not actually improve their ability to taste, but their enjoyment of tastes may increase dramatically.

Tart's (1971) sample revealed a related, characteristic effect: intoxicated individuals enjoyed eating and reported consuming large quantities of food. They also commonly craved sweets during intoxication. Both of these effects began at low levels of intoxication. In a separate sample of 100 Caucasians who had smoked cannabis at least 50 times, 72% said that the drug usually increased their hunger and 37% said it increased their desire for sweets (Halikas et al., 1971).

A detailed laboratory study confirmed these reports. This research revealed a 40% increase in calorie consumption during intoxication. The study had six men live in a laboratory setting for 13 days. Each day they smoked 4 marijuana cigarettes or 4 placebos. Not only did they consume more calories on the days that they smoked cannabis, but they also gained more weight than one would predict from these additional calories. This result suggests that marijuana may slow metabolism as well as increase food consumption (Foltin, Fischman, & Byrne, 1988). Results like these have inspired the medical use of cannabinoids to improve appetite for people with problematic weight loss, as discussed in chapter 8.

Higher Functions

Emotion

Any drug's impact on human feelings determines its potential for repeated use. Literary works devoted to cannabis frequently mention its pleasant influence on emotion. Tart's (1971) sample reported that cannabis almost invariably improved their mood. This effect appeared at moderate levels of intoxication or more. Users also grew more relaxed at this level of intoxication. Data from another sample of 100 people who used the drug at least 50 times revealed consistent reports of peaceful and relaxed feelings after smoking (Weller & Halikas, 1982). More than 2,500 veterans who had smoked at least 5 times also reported many pleasant effects of cannabis. More than 90% said that the drug made them feel mellow or relaxed. Over 60% reported that the drug made them euphoric (Lyons et al. 1997). These reactions likely motivated continued consumption of the drug.

These emotional effects of cannabis are not only pharmacological but also may stem partly from expectancies. Evidence for the role of expectancies in cannabis's emotional impact comes from laboratory research. For example, people who expect to smoke hashish in the laboratory report feeling "high," even if the hash contains no THC (Cami, Guerra, Ugena, Segura, & De La Torre, 1991). Thus, part of the emotional impact of the drug arises in the user's own mind.

In addition, the idea that the drug's effect is actually pleasant may depend upon the user's attitude. A study using synthetic THC gave the drug to two groups of people with different instructions. One group knew the drug was THC; the other group only knew that the drug was an antiemetic. People who knew that the drug was THC liked the effects more, found them more euphoric, and wanted more of the substance. People who did not know that the drug was THC were significantly less positive about it. Thus, expectations about marijuana and its effects likely contribute a great deal to its emotional impact (Kirk, Doty, & de Wit, 1998).

Tart's (1971) work documents other affective reactions, too. His sample reported that they commonly felt emotions more strongly after using cannabis. This effect did not usually begin until participants reached strong levels of intoxication. Some examples in literary works support this idea, but laboratory studies have yet to address the question. Several methods for assessing emotional reactions have developed over the years. A simple study comparing those who smoked cannabis to those who smoked placebo might elicit reactions to emotional slides or film clips. Greater reactions in the cannabis users would support this report of exaggerated emotions.

Tart (1971) also investigated emotional crises during marijuana intoxication. He used the jargon of the era, asking participants the percentage of users whom they had seen "freak out" or feel "catastrophic emotional upset." The vast majority of the sample (89%) estimated that this effect occurred less than 1% of the time. The actual rate of aversive reactions to marijuana is probably higher than the number reported by this sample of experienced users who clearly enjoy the drug.

Thought

Cannabis's impact on emotion may relate to some of its effects on thinking. Many of the drug's cognitive effects appear in chapter 4. Users report a number of subjective impressions about these changes in their thoughts. Tart's (1971) sample reported several relevant, characteristic effects. At strong levels of intoxication or more, they felt that their thoughts were more "in the present" or "here and now." At levels of intoxication from fairly strong or higher, they found that they were more likely to make spontaneous insights about themselves, appreciate subtle humor, and accept contradictory ideas. No laboratory studies have addressed these effects directly.

Tart's (1971) participants also reported trouble reading when they were this intoxicated. In contrast, a separate sample of 100 regular users found that 30% reported usually experiencing better concentration and improved mental powers during intoxication (Halikas et al., 1971). Laboratory studies generally contradict these impressions of cognitive improvement during intoxication. Perhaps the drug creates the illusion of improved concentration despite deficits.

Memory

Marijuana alters some aspects of memory, as documented in chapter 4. The subjective experience of memory may differ markedly from the actual ability. The subjective experience parallels many of the laboratory studies. Users rarely report problems remembering material learned prior to intoxication. Laboratory studies generally confirm that people can remember old material while high. In contrast, users do report deficits in short-term memory during intoxication. Tart's (1971) participants characteristically forgot the topic of conversations even before they had ended. More than half of the sample stated that this forgetting of conversations occurred very often or usually. This effect began at strong or very strong levels of intoxication, as laboratory studies of memory confirm.

Although not a characteristic effect, a more dramatic impairment of memory appeared commonly in Tart's (1971) data. Over 65% of the sample said that at least some of the time when they were intoxicated they could not remember the beginning of a sentence by the time they reached its end. This drastic impairment of short-term memory also began at strong or very strong levels of intoxication. This sort of forgetting appeared commonly in reports from more than 200 Canadian users, too (Adamec et al., 1976). Users obviously have some insight into the memory deficits that appear soon after cannabis consumption.

Tart's (1971) data also revealed an intriguing and unexpected effect related to memory. Users commonly reported that they spontaneously recalled events from the distant past, including material they had not considered in many years. For example, people might recall an incident from grade school that they had not thought about for quite some time.

This effect began at strong levels of intoxication for the majority of the sample. Users appear to know that their short-term memory suffers after smoking cannabis, but they also claim that spontaneous recall of distant memories improves.

Sexuality

Few topics are more controversial in American society than sex and drugs. Their combination often generates confusion and concern. Marijuana's link to sex may be as old as the drug itself. As with other effects, this one first appeared in literature. One of the tales in *The Arabian Nights* (*1,001 Nights*), published and popularized by 1200 B.C., mentions sexual arousal in a man who has eaten hashish. Louisa May Alcott's (1869) short story "Perilous Play" suggests hashish may speed seduction. Harry Anslinger spun tales of cannabis enhancing sexuality in his efforts to pass the Marijuana Tax Act of 1937. These reports relied on only a few cases. Larger studies confirmed the belief that marijuana alters aspects of sexuality.

The most characteristic effect related to sex for Tart's (1971) participants concerned enhanced orgasm. Users reported that they appreciated new qualities of orgasm that they did not usually experience when sober. This effect may parallel a general increase in the excitement, joy, and sensitivity of touch, which was also characteristic of intoxication in this sample. Over half of the participants reported that they were better lovers after using the drug, with many suggesting that they were more responsive and giving. Most of these effects did not begin until at least a moderate degree of intoxication.

Self-report research on a separate group of 100 experienced users confirmed marijuana's impact on sex (Weller & Halikas, 1984). Two-thirds of this sample, who had used the drug at least 50 times, reported that cannabis intoxication led to some form of sexual enhancement. They reported improved orgasm, a heightened sense of intimacy and closeness, and superior sexual prowess. Coincidentally, these users stayed single longer and were more likely to have sexual contact with someone of their same sex than people who did not use the drug. Although many effects of marijuana can dissipate over time, marijuana's enhancement of sex appears to remain stable across 6 to 8 years (Halikas, Weller, Morse, & Hoffmann, 1985).

Few laboratory studies have confirmed these self-reports. Studies of these sexual effects might include masturbation or intercourse after the administration of THC. Research of this type could validate reports of enhanced sexual experiences during intoxication. This work might suggest a new treatment for some sexual dysfunctions. A relatively common and important problem, hypoactive sexual desire disorder, might benefit from marijuana. The hallmark symptom of this disorder is an extremely low sex drive. A decreased desire for sex commonly arises from medical or psychiatric conditions as well as poor relationships. Once these potential causes have been eliminated, marijuana may prove a fruitful way to increase sexual desire. Despite this potential promise, studies of cannabis's impact on sexual drives have not been a high priority of most research funding agencies.

Spirituality

Another controversial topic in American culture concerns concepts related to the divine. Scientific research on the holy, religious, sacred, or spiritual often offends some people. Empirical approaches to these topics were taboo for many years. Nevertheless, recent research documents that spirituality provides superb benefits for mental and physical health (Miller, 1999). These results are hardly news to many people leading religious lives. Yet adding illicit drugs into this sort of research remains controversial.

Several cultures view psychoactive substances as an important part of spirituality. For example, the Native American Church uses peyote as a sacrament. The Coptic and Rastafarian Churches smoke cannabis as part of their religious practice, too. Certain sects of Buddhism in Nepal use marijuana as a sacrament (Clarke, 1998). Thus, spiritual aspects of cannabis have inspired some investigation.

Tart's (1971) sample reported only one characteristic effect that he interpreted as potentially spiritual. This effect concerned feeling more childlike, open to experience, and filled with wonder. Over 65% of the sample experienced this effect very often or usually. It began at moderate to strong levels of intoxication. Tart (1971) also asked simple yes-or-no questions about spiritual topics. One-fourth of the sample reported spiritual experiences from marijuana that had a dramatic impact on them. Users described these events as moments of connection to the universe, contact with the divine, or expressions of peace and joy. These effects paralleled reports of religious ecstasy. Approximately one-fifth of the sample said that intoxication had acquired religious significance for them. Contemporary authors also assert that the drug can enhance spirituality. Many encourage pensive, meditative use of the drug and deride mindless consumption (Bello, 1996). This approach to use may minimize the potential for negative consequences related to the drug. People who smoke cannabis in a thoughtful way and consciously attend to their experience may be less likely to show symptoms of abuse.

Sleep

Marijuana intoxication alters sleep. Dr. J. R. Reynolds, chief physician to Queen Victoria, recommended the drug for insomnia. Many early literary accounts mention sedation and dramatic dreams (Rosenthal, Gieringer, & Mikuriya, 1997). Tart's (1971) sample commonly reported that they grew drowsy, particularly at strong levels of intoxication. They characteristically stated that they found falling asleep very easy, beginning at the lowest level of intoxication. They also reported improved sleep quality, especially at strong levels of intoxication.

On the other hand, a subset reported disturbed sleep, especially after very high doses. This paradoxical arousal goes against other self-report studies that confirm that marijuana relaxes people (Lyons et al., 1997; Halikas et al., 1985). Laboratory research has revealed greater sedation when participants smoke cannabis. The placebo joint did not have the same effect (Block, Erwin, Farinpour, & Braverman, 1998). These effects have inspired cannabis use in the informal treatment of insomnia. Many other drugs have an impact on sleep, particularly the barbiturates and benzodiazepines. The barbiturates are notorious for their potential for abuse, dependence, and lethal overdose. Benzodiazepines can cause memory loss and lead to a sluggish feeling the next morning.

The drawbacks of these insomnia drugs led a woman with multiple sclerosis to smoke marijuana before bed. She reported successful, restful sleep as a result (Grinspoon & Bakalar, 1997). Although THC causes many of marijuana's effects, cannabidiol appears to have the biggest impact on sleep. A sample of 15 insomniacs who received cannabidiol improved their sleep dramatically (Carlini & Cunha, 1981). Despite these encouraging data for cannabidiol, some of the best treatments for insomnia require changing behaviors rather than taking drugs. These interventions include multiple steps. People with sleep problems often benefit from retiring at the same time each night, avoiding stimulants like caffeine, and using their beds only for sleep and sex rather than other activities. This sort of good sleep hygiene may provide better rest than any medications. Nevertheless, further research on smoked marijuana and isolated cannabidiol can provide intriguing information on the role of the cannabinoids in sleep and consciousness.

Undesirable Effects

Negative feelings associated with marijuana intoxication often receive less attention than the stereotypical euphoria. Cannabis can create aversive reactions, particularly after extremely large doses or during the first exposure to the drug. Literature has not neglected the distressing impact hashish may have. Gautier, Ludlow, and Baudelaire all detail frightening effects associated with overdose. One of Louisa May Alcott's (1869) characters in "Perilous Play" describes the aversive effects as "not so pleasant, unless one likes phantoms, frenzies, and a touch of nightmare." Tart's (1971) sample did not report many negative reactions. They claimed that they often found themselves distractible and easily sidetracked. This mental fogginess was the only characteristic negative effect. Common negative effects included an inability to think clearly, work accurately, or solve problems efficiently. Participants also said that marijuana made them feel physically weaker.

Laboratory research confirms slow and inefficient thought during intoxication. Experiments have not documented physical weakness, but reported sedation in the laboratory may reflect this feeling (Block et al., 1998). Tart's (1971) minimal reports of negative consequences like panic or discomfort may not be typical of everyone. His participants had all smoked cannabis an average of over 200 times, with a minimum of a dozen. People who experience severe negative reactions likely quit using the drug long before the twelfth try. Therefore, they would not end up participating in studies requiring consistent marijuana consumption.

A sample of more than 2,500 people who had used cannabis at least 5 times confirmed these negative effects and suggested a few more. This study may have revealed more negative effects because it did not require as much use of the drug as Tart's (1971) research. Over half of the sample claimed that they could not concentrate when they were intoxicated, and nearly 40% said that the drug made them confused. Participants also reported many other undesirable reactions, including paranoia, guilt, and nausea. Some of the people in this study were twins, permitting an examination of the heritability of these effects. Analyses comparing the identical twins to the fraternal twins revealed that these negative effects were likely inherited. Positive effects, which included enhanced relaxation, creativity, energy, and euphoria, also appeared to have a heritable component (Lyons et al., 1997). These results support the idea that a biological factor contributes to cannabis's subjective effects.

Another potentially negative feeling associated with marijuana intoxication is depersonalization. Depersonalization typically involves an alteration in the experience of one's self or reality. Feeling unreal, separated from one's body, or anxiously unaware of identity is part of depersonalization. It can occur during a number of unfavorable conditions, including sleep deprivation, fatigue, panic, and psychosis. Nevertheless, under appropriate circumstances the sensation may not feel aversive. Sensory deprivation and meditation may lead to depersonalization with few frightening or disorienting qualities. Tart (1971) did not inquire about this effect. A study of 100 regular users found 12% reported usually feeling a separation from self after smoking marijuana. Almost half of this sample (49%) said they have had this experience occasionally (Weller & Halikas, 1982). Laboratory work clearly documents that cannabis heightens depersonalization (Mathew et al., 1999). This depersonalization correlated with anger, tension, and confusion, suggesting that the experience had negative components.

Two other undesirable effects of marijuana include dry mouth and red eyes. Over 60% of a sample of 100 experienced users reported that smoking marijuana usually dried their mouths and throats. Almost all of the sample (99%) experienced this effect at least occasionally. Two-thirds said that marijuana made their eyes red at least occasionally (Halikas et al., 1971). Users easily cure dry mouth with a few sips of liquid, and red eyes usually respond to drops. Thus, these negative effects are not strong deterrents to consumption of the drug. Most users complain about red eyes as a telltale sign of intoxication that they would prefer to avoid in many settings.

One of the most novel and striking undesirable reactions to cannabis illustrates the role of cultural factors in drug responses. At least two individuals who smoked high doses of marijuana the first time that they tried the drug experienced Koro (Chowdhury & Bera, 1994), which means "turtle's head." It is an acute state of anxiety associated with a strong fear of death. It also includes the alarming perception that one's penis has retracted into the abdomen. Any man who holds his member dear can understand the terror that must accompany this delusion. Thankfully, the disorder remains extremely rare. Oddly, most cases are limited to Asian countries, where the idea that anxiety might lead to penis loss is considered more tenable. In China, the disorder is known as *shook yang* (shrinking penis). The two cases associated with marijuana intoxication appeared in West Bengal, India, where a Koro epidemic had occurred in 1982 (Franzini & Grossberg, 1995).

The first reported case of marijuana-induced Koro involved a 27-year-old Hindu, who took 30 large inhalations of cannabis the first time he used the drug. He later had odd sensations in his legs and reached down to touch them. To his horror, he found that his penis had seemingly disappeared inside his abdomen. Understandably, he screamed for help. His friends came, grabbed his penis, and put him in a nearby pond for over two hours until he realized his genitals were normal. The second case involved a distant cousin of the first. This 26-year-old man lived in a nearby village. He had heard of his cousin's case but did not know that the symptoms appeared after using cannabis. On his first exposure to the drug he felt an odd, empty space in his abdomen that he thought he could fill via deep breathing. He suddenly had the haunting sensation that each breath caused further and further retraction of his penis. He, too, cried for help. His friends placed him in a pond for half an hour until he felt his genitals had returned to normal.

Explanations of this odd malady remain difficult to prove. The phenomenon remains so rare that systematic studies have been impossible. Nevertheless, given the documented increases in anxiety that can accompany marijuana intoxication, these cases may represent a cultural interpretation of panic. Both men lived near the location of a Koro epidemic. They may have learned vicariously that panic and penile retraction can occur together. Thus, in their first exposure to cannabis, anxiety and other symptoms may have led them to think of this reaction to panic. Once this expectation was activated, the reaction may have become self-fulfilling. Worry about penile retraction may have exaggerated anxiety, which in turn may have heightened the worry about penile retraction. The simple treatment (spending time in a pond) suggests that perhaps any distraction that might alleviate anxiety could decrease Koro. Other cases support the idea that anxiety and a cultural expectation contribute to the disorder. For example, Koro has appeared during heroin withdrawal, a condition notoriously associated with angst and discomfort

(Chowdhury & Bagchi, 1993). Laboratory investigations of marijuana-induced Koro have not appeared in the literature.

Another undesirable side effect worthy of investigation concerns cannabis hangover. Clinical lore suggests that a night of heavy marijuana consumption can lead to fuzzy thinking and fatigue the following day. Self-report questionnaires offer mixed results about reactions that linger after intoxication has ended. Tart's (1971) seminal work on subjective effects did not address hangover.

A study of 100 people who had smoked at least 50 times did address this question. These people rated many possible aftereffects as occurring "usually," "occasionally," or "not at all." The aftereffects that most of the sample rated as "usual" ones were all positive. More than half of the participants said that they usually experienced feeling calm, clear-minded, and rested after marijuana intoxication wore off. In contrast, approximately half also reported that occasionally they awoke tired and felt that their minds were foggy. These experiences are more consistent with the stereotype of hangover.

These self-report data suggest that the aftereffects of marijuana do not feel as aversive as the hangover symptoms associated with alcohol or other drugs. In fact, it is unclear from these data if marijuana is the actual source of all of these symptoms, given the frequency that people consume alcohol, marijuana, and other drugs simultaneously (Earleywine & Newcomb, 1997). Controlled administration of marijuana in the absence of other drugs is the only way to illuminate this issue.

Laboratory studies do not consistently confirm the presence of a marijuana hangover. Subjective experiences of hangover probably vary as dramatically as reactions to intoxication. One study of 12 subjects found no evidence of hangover the morning after smoking cannabis in the laboratory (Chait, 1990). Another experiment performed in the same laboratory is frequently cited as evidence for hangover because participants who had smoked cannabis felt worse the morning after. Yet a close look at the results reveals that people felt significantly better the morning after smoking cannabis than after smoking placebo. The 13 participants scored higher on Elation and Positive Mood scales in the cannabis condition. They also reported feeling more energetic and aroused. Perhaps participants fell into a bad mood when they expected to smoke marijuana and ended up with placebo. This disappointment may have stayed with them until the next morning (Chait, Fischman, & Schuster, 1985).

Other studies offer more support for a marijuana hangover. Researchers administered 10, 20, or 30 mg of THC. (These dosages translate to roughly the amount of THC in .5, 1, and 1.5 cannabis cigarettes. Given the typical loss of 50% of THC to sidestream smoke, the impact may have been more like 1, 2, and 3 joints.) The 9 participants in this study reported some residual intoxication and confusion the next day, particularly at the highest dose (Cousens & DiMascio, 1973). Perhaps THC alone causes more negative aftereffects than the full combination of cannabinoids present in marijuana.

Behavioral measures also suggest marijuana may have some impact after intoxication has ended. A distorted perception of time can remain the morning after smoking marijuana (Chait et al., 1985). A study of nine airplane pilots showed an unsurprising impairment on a flight simulator after smoking one cannabis cigarette. In seven of them, performance did not return to unintoxicated levels, even 24 hours later (Leirer, Yesavage, & Morrow, 1991). Thus, though marijuana hangover lacks the severity of the aftereffects of alcohol and other drugs, it can occur at detectable levels for laboratory study. These negative experiences associated with use may alter consumption of the drug in many people.

Conclusions

It's hard to describe consciousness. Anyone's subjective experience always includes complicated combinations of thoughts, feelings, and sensations. The way these combinations alter during marijuana intoxication remains difficult to depict. Nevertheless, literary examples, case studies, laboratory experiments, and reports from experienced users confirm several of cannabis's effects. Although individual reactions vary dramatically, a few key experiences appear commonly in regular users. The drug clearly alters perception. Time slows. Space appears more vast or variable. The senses generally seem more appealing and interesting, despite laboratory evidence that they may actually be impaired. Visual acuity seems better. Sounds appear to take on new qualities. Touch and taste both seem more intriguing and sensual. Yet laboratory evidence does not support these enhancements.

Higher functions also change during marijuana intoxication. Emotions seem more salient or extreme. Euphoria predominates. Thoughts seem more focused on the current moment. Short-term memory clearly suffers, with users occasionally forgetting one sentence while uttering the next. Sexuality and spirituality increase. Sleep can improve at low doses or suffer at higher ones. A few negative subjective effects also seem common, including anxiety, guilt, paranoia, and perhaps hangover. Some of these effects may stem simply from expectancy, some vary with culture, and some clearly arise as part of the pharmacology of the cannabinoids. The drug's popularity may rely, in part, on its ability to create all of these disparate but potentially pleasant effects.

6

Cannabis Pharmacology

Understanding marijuana's impact on biological systems requires a description of its active components. This chapter begins by identifying the mind-altering molecules in marijuana, the substances that contain these chemicals, and their respective potencies. It continues with a discussion of the way these substances enter the body, metabolize, and reach their sites of action. The remainder of the chapter focuses on the receptors that respond to these cannabinoids and the natural substances in the body that work at these same sites.

Marijuana contains more than 60 compounds unique to the plant called cannabinoids. They interact with each other in interesting ways, altering their impact. Cannabinoids appear in a variety of strengths in marijuana, hashish, hash oil, and synthetic medications like nabilone, dronabinol, and levonantradol. People eat or smoke these products, leading to slower or faster absorption of chemicals. The cannabinoids alter the permeability of nerve membranes. They also react with their own special receptors—CB1 in the brain and nervous system and CB2 in the immune system. Researchers have identified substances native to the body that also work on these receptors, including anandamide and arachidonolyl-glycerol. Details of each of these topics appear in the following sections.

Active Ingredients—the Cannabinoids

A first step in understanding marijuana's impact involves identifying its active components. Cannabis contains more than 400 different chemical

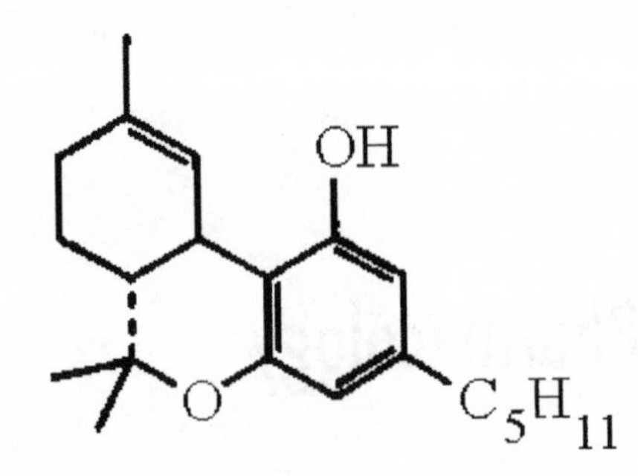

Figure 6.1. Delta-9-THC. This cannabinoid is responsible for most of marijuana's psychoactive effects.

compounds. At least 66 are unique to the plant and receive the name "cannabinoids." The best known cannabinoid is probably delta-9 tetrahydrocannabinol (THC). Considerable research has also examined the related molecule delta-8 THC. Two conventions exist for naming chemicals: formal and monoterpenoid. Thus, delta-9 THC (the formal name) is also called delta-1 THC (the monoterpenoid name). Similarly, delta-8 THC is also known as delta-6 THC. This text uses only the formal names. THC alone refers to the delta-9 variety. Delta-9 THC and delta-8 THC appear to produce the majority of the psychoactive effects of marijuana. As figures 6.1 and 6.2 reveal, the molecules differ only in the location of the double bond in the first carbon ring.

Delta-9 THC is more abundant in the plant, leading researchers to hypothesize that it is the main source of the drug's impact. The liver breaks delta-9 THC down into 11-OH-delta-9 THC (11-hydroxy-delta-

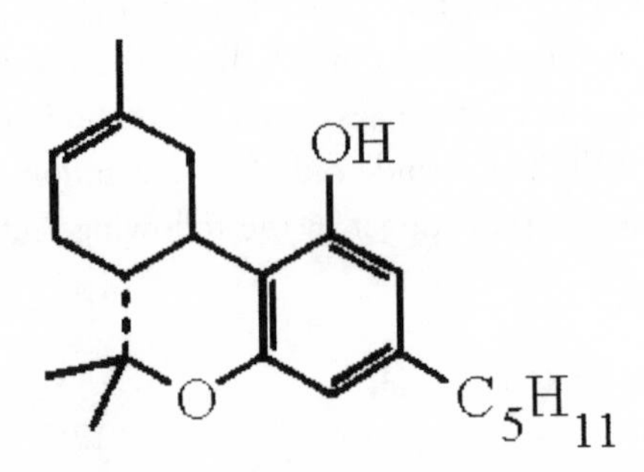

Figure 6.2. Delta-8-THC. This cannabinoid produces some of marijuana's psychoactive effects, but it is less abundant than delta-9-THC.

9-THC). This metabolite also causes psychoactive effects, including changes in subjective sensations. As figure 6.3 reveals, it only differs from THC by a few atoms, but it may be three times as potent because it reaches the brain more readily (Razdan, 1986).

Two other common cannabinoids are cannabinol and cannabidiol, depicted in figures 6.4 and 6.5, respectively. Delta-9-THC, cannabinol, and cannabidiol are the most prevalent psychoactive chemicals in the plant and provide the majority of marijuana's effects. For example, THC and cannabidiol account for 95% of marijuana's active ingredients (Doorenbos, Fetterman, Quimby, & Turner, 1971). Dozens of other cannabinoids exist, but most are variants of delta-9 THC, delta-8 THC, cannabinol, and cannabidiol. Research has uncovered six additional families of molecules unique to marijuana. All begin with the familiar "cannab" prefix. These include cannabichromene, cannabicyclol, cannabielsoin, cannabigerol, cannabinidiol, and cannabitriol. Many differ little from each other. All are lipophilic, meaning that they dissolve in fat, fatty tissue, or fatty fluids. They are not soluble in water. Thus, despite the claims of many aging hippies, teas made from boiled marijuana probably do not create extensive cannabinoid effects.

A great deal of research focuses on THC. Some investigators have turned their attention to the other cannabinoids, particularly cannabinol and cannabidiol. Studies address the activity of each of these chemicals alone and in combination with THC. Cannabinol has generated considerable interest, in part, because THC breaks down into this compound

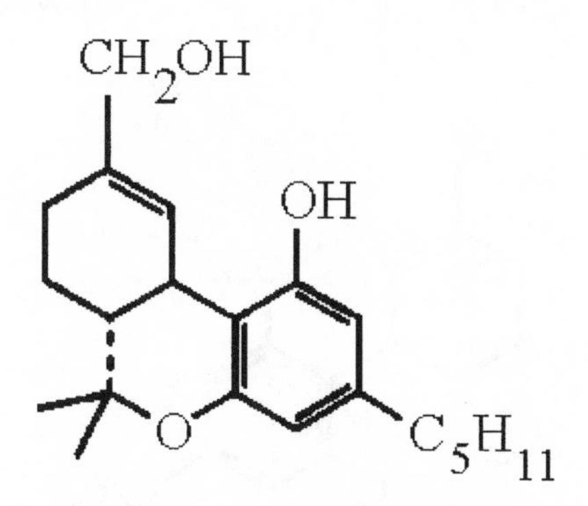

Figure 6.3. 11-hydroxy-delta-9-THC. The liver breaks delta-9-THC into this compound, which reaches the brain faster and may be 3 times as psychoactive.

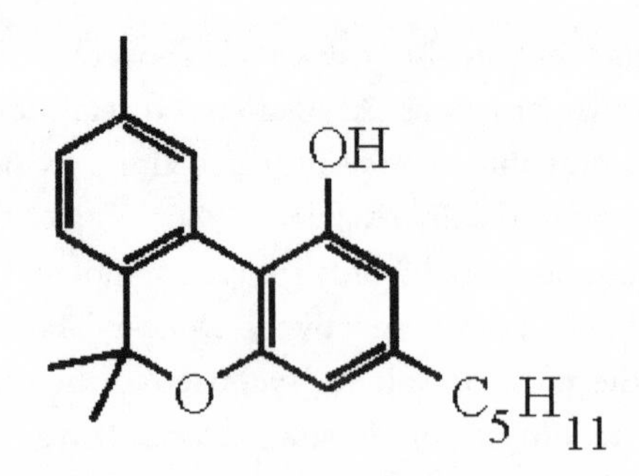

Figure 6.4. Cannabinol. Delta-9-THC breaks down into this compound as stored marijuana ages.

as it ages. Initial research on cannabinol suggested that it had no impact on subjective experience, offering an explanation for the decreased potency of old cannabis (Hollister, 1974). As the THC in marijuana degrades into cannabinol, the marijuana's effects diminish. Yet intravenous administration of cannabinol can create some subjective effects at high doses, since it appears to be about one-tenth as strong as THC (Perez-Reyes, Timmons, Davis, & Wall, 1973).

Animals clearly respond to high doses of cannabinol as if it were comparable to THC (Jarbe & Hiltunen, 1987). For example, both cannabinol and THC increase sleep and lower body temperature in mice (Yoshida et al., 1995). Despite these similarities, cannabinol is probably more active in the immune system than in the nervous system. In contrast, THC is probably more active in the nervous system than in the immune system (IOM, 1999). Because both compounds are so prevalent in marijuana,

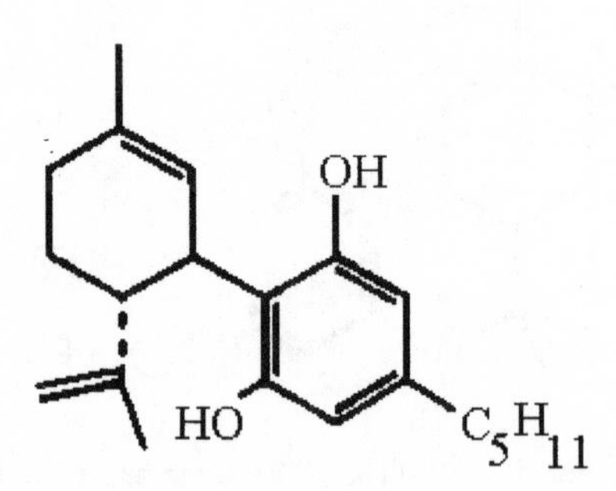

Figure 6.5. Cannabidiol. This cannabinoid becomes delta-9-THC as the marijuana plant matures.

research has also addressed cannabinol's interaction with THC. Cannabinol diminishes the intensity of THC's subjective effects but prolongs their duration. Cannabinol may also decrease some of the stimulation that THC can produce (Brazis & Mathre, 1997).

Considerable research also focuses on cannabidiol. Cannabidiol becomes THC as the marijuana plant matures, and this THC later breaks down into cannabinol. Up to 40% of the cannabis resin from some marijuana strains is cannabidiol (Grilly, 1998). Yet the amount varies in different plants. Some African varieties contain little of the chemical (Turner & Hadley, 1974). Early studies suggested that cannabidiol administered alone had no impact, much like early studies of cannabinol (Hollister, 1974). Later research, however, revealed its effects only appeared at moderate doses. Too little or too much of the drug created no response. The appropriate dosage, however, can decrease anxiety in healthy people. It also reduces psychotic symptoms such as hearing voices or thinking incoherently. In addition, cannabidiol induces sleepiness and may protect epileptics against seizures (Zuardi & Guimaraes, 1997).

Research has also addressed the impact of cannabidiol on THC's effects. The two cannabinoids in combination may create different experiences from either drug alone. Synergistic interactions like this can be very difficult to study. Drug lore suggests that cannabidiol minimizes THC's psychoactive effects and delays their onset. Yet formal studies reveal that the interaction is not quite so simple. Cannabidiol may exaggerate some of THC's effects while attenuating others. It may increase THC-induced euphoria, but limit the production of anxiety and disordered thinking. Cannabidiol slows THC metabolism in the liver. Thus, a dose of THC combined with cannabidiol will create more psychoactive metabolites than the same dose of THC administered alone (Bornheim, Kim, Perotti, & Benet, 1995). By slowing THC's metabolism, cannabidiol can exaggerate some of its effects, including euphoria and the subjective sense of feeling high.

Although cannabidiol slows THC metabolism, it also may limit the drug's negative side effects. THC alone can produce anxiety, panic, and psychotic symptoms, particularly at high doses. Cannabidiol not only decreases anxiety on its own, but it also buffers against THC-induced panic and discomfort. Cannabidiol minimizes psychotic symptoms like bizarre thoughts or odd perceptions. Thus, it may attenuate these negative aspects of THC intoxication. Strains of marijuana that lack cannabidiol may produce more panic or psychotic effects. These findings may prove particularly useful given recent research on dronabinol (Marinol), the synthetic version of THC used to treat nausea and weight loss. Negative side effects of this drug might decrease if physicians combined it with cannabidiol (Zuardi & Guimaraes, 1997).

Cannabis Preparations

Users can ingest cannabinoids in a number of forms, including marijuana, hashish, hash oil, and synthetic medications. Nearly all parts of the marijuana plant contain psychoactive ingredients, but most of the cannabinoids appear in the resinous glands and flowering tops. Thus, the price for glands or tops is markedly higher than for other parts of the plant. Different cannabis preparations have different names. "Marijuana," a Spanish word purportedly coined in Mexico, originally meant cheap tobacco. The term may stem from the Portuguese expression "mariguango," which means intoxicant (Maisto et al., 1995). Later the word referred to the dried leaves and flowers of cannabis. Residents of India distinguish among three forms: bhang, ganja, and charas. Bhang is the dried leaves of the plant, comparable to marijuana. People smoke these leaves or combine them with milk and spices to form a drink that is also called bhang. Ganja refers to the sap-carrying tops of female plants in India, but in Jamaica the term applies to the leaves as well. Charas is hashish, the dried resin separated from the flowers and pressed together (McKim, 1997).

Many legends surround hashish. A well-known report concerns an exotic technique for collecting resin. Harvesters allegedly pranced naked through sunny fields of cannabis, gathering shiny sap and scraping it gently from their bodies to form cakes. These tales sound like contemporary urban legends or a marketing strategy for modern dealers. Nevertheless, comparable stories appeared as early as the 1850s. Most reporters from that era found workers who wore leather aprons to catch resin as they ran through fields. These workers then told of naked harvesting in other locations (Johnston, 1855; Von Bibra, 1855).

Stories like these may have been an attempt to fool outsiders, similar to the way children from farms tell children from the city that chasing a cow will turn her milk to cottage cheese. Perhaps poorer gatherers of hashish resin could not afford leather aprons. Whatever the arrangement in the past, modern hashish production does not employ naked trips through fields. Instead, manufacturers shake the resinous glands from the plants and press it into hash. Others simply form blocks by pressing hash oil into powdered cannabis.

In addition to ganja, bhang, and charas or hashish, hash oil also appears in the illicit drug market. Producers create this viscous liquid by boiling hashish or cannabis in a solvent, straining it through a filter, and then letting the solvent evaporate, leaving the oil. The process can be extremely dangerous given the flammability of the solvents, which usually include alcohol or ether (Gold, 1989). The risk may prove worthwhile because hash oil commands higher prices. The oil has the potential to generate huge profits because it is relatively compact and easy to smuggle. It is often more potent than hashish or cannabis, too. Yet hash oil is not particularly popular. Smoking the oil by itself can require special glass pipes as part of a messy and cumbersome process. The solvents used to form the oil may be unhealthy to smoke. Most users find hashish or cannabis easier to ingest and potent enough to create the effects they desire (Clarke, 1998).

Synthetic cannabinoids also exist; they are usually ingested orally. Dronabinol (Marinol), a synthetic version of THC suspended in sesame oil, can treat poor appetite, nausea, and vomiting (e.g., Lefkowtiz et al., 1995). Current studies also address dronabinol's efficacy as a treatment for spasticity associated with multiple sclerosis and pain after surgery (Hanigan, Destree, & Truong, 1986). Some initial work also suggests that this drug might help disturbed behavior in Alzheimer's patients (Volicer et al., 1997). The abuse potential for this substance appears to be minimal. It is the only cannabinoid approved for medical use in the United States. Nabilone (Cesamet), an analogue of THC available in the United Kingdom, also limits nausea, vomiting, and spasticity (e.g., Steel et al., 1980). Levonantradol, another synthetic THC analogue unavailable in the United States, shows some promise in treating acute surgical pain, nausea, and vomiting (Jain et al., 1981; Tyson et al., 1985). Apparently, no pharmaceutical company has pursued its development as a medication (IOM, 1999). The medicinal uses of these synthetic cannabinoids appear in more detail in chapter 8.

Potency

Cannabis preparations vary dramatically in their effects. The most common indicator of potency is the percentage of delta-9 tetrahydrocannabinol (THC). Although THC is not the only source of psychoactive effects in the plant, it is the most abundant chemical that clearly alters subjective experience. Hashish typically contains 20% THC, with some estimates as high as 50%. Hash oil can contain up to 70% THC. Yet each of these products can vary dramatically in potency. Some samples of hash oil and hashish contain no THC at all. These products with no THC obviously create few subjective effects, except for those that arise from expectancy. Hashish and hash oil with the highest potencies often cause the most dramatic experiences of intoxication.

Marijuana also shows considerable variation in potency depending upon the plant strain, growing conditions, and storage. Some varieties of plants contain more THC than others. *Cannabis sativa* used for industrial hemp often contains less than 1% THC. Smoking marijuana this low in potency does not change subjective experience. Marijuana with less than 1% THC has the same effects as a placebo (Zimmer & Morgan, 1997). Thus, hemp products are not psychoactive. No one will grow intoxicated from smoking the various shampoos, soaps, or clothes currently manufactured from these plants. Psychoactive strains of marijuana typically contain 2 to 5% THC, but concentrations as high as 22% have been documented (Iversen, 2000). The moisture and temperature of the growing season can alter potency. Storage in hot environments can degrade the cannabinoids and lower THC content (Clarke, 1998). Exposure to light also accelerates the breakdown of THC. A year of storage in a bright place can produce nearly three times the decrease in THC as a year of storage in a dark place (Brazis & Mathre, 1997).

Many media reports suggest that cannabis has increased in potency quite dramatically in recent years. These reports have generated considerable debate. Yet the magnitude of the increase is difficult to document. In addition, the tacit assumption that increased potency translates into greater danger from the drug may not be true. Reports of a stronger drug actually began over 30 years ago. By the middle of the 1980s, some authors suggested that marijuana's potency had increased by a factor of 100 (MacDonald, 1984). These claims clearly suffered from exaggeration or misinformation. Other arguments about increased potency arose from the University of Mississippi's Potency Monitoring Project. This program reports the average THC content of cannabis taken in drug arrests. Estimates were extremely low in the 1970s, sometimes below 1%. As discussed above, cannabis with this little THC has no impact on subjective experience. The idea that a drug with no effects would increase in popularity over the years makes little sense. Thus, these estimates from the 1970s were probably poor reflections of the amount of THC in marijuana available at the time.

Investigators hypothesize that the data from the Potency Monitoring Project underestimate the true amount of THC in marijuana from the 1970s. First, the estimates were based on very few samples of seized cannabis. In some years there were no more than 50 samples to analyze (Potency Monitoring Project [PMP], 1974–1996). In addition, police may have stored the marijuana in hot lockers that allowed the THC to degrade rapidly (Mikuriya & Aldrich, 1988). Despite the small samples and poor storage, the average THC content in 1976 was 2% (ElSohly, Holley, & Turner, 1985).

An alternative source of potency information, an independent laboratory in California, analyzed many more samples than the Potency Monitoring Project and found a large range in THC concentration. In 1973, this laboratory tested over 100 samples and found that marijuana had an average THC content of 1.6 % (Ratcliffe, 1974). Later analyses ranged up to almost 8% THC (Perry, 1977). Thus, the idea that all cannabis of the 1970s had less than 1% THC seems unlikely. Ratcliffe's (1974) estimate of 1.6% may be conservative but credible; the 1976 estimate of 2% may be closer to the truth.

Potency data from the 1980s through the middle of the 1990s suggest that THC content continued to vary dramatically from strain to strain and sample to sample. With improved storage techniques and much larger samples, the Potency Monitoring Project found THC concentrations varied from 2% to almost 4%. Average concentrations approached 4% THC in 1984, 1988, 1990, and 1991 (PMP, 1974–1996). Trends in the rest of the 1990s showed comparable THC content, with a peak around 4.5% THC in 1997. Other cannabinoids like cannabinol and cannabidiol have not increased in concentration over the years (ElSohly et al., 2000). Thus, claims of 1,000% (Cohen, 1979) or 10,000% (MacDonald, 1984) increases in marijuana potency are clearly inaccurate. A threefold increase from approximately 1.5% in the early 1970s to 4.5% in the late 1990s may be closer to the truth. A simple doubling from an average of 2% to an average of 4% also seems plausible.

Although many media reports warn that increased potency translates into greater danger, data suggest otherwise. The implications of a two or threefold increase in THC concentration remain unclear. Marijuana with greater amounts of THC may not prove more hazardous than weaker cannabis. First, acute administration of the drug is essentially nontoxic. No one has ever died from THC poisoning. Smoking enough cannabis to ingest a lethal amount of THC may be physically, if not financially, impossible.

Estimates of a fatal dose of any drug arise from some rather gruesome animal research. Different groups of animals receive large amounts of a drug until a particular dosage kills 50% of them. Researchers refer to the dose that is lethal for 50% of the animals as the LD 50. Investigators then extrapolate from these data to estimate a lethal dose for humans. The LD 50 for THC is approximately 125 mg for every kilogram of body weight (Nahas, 1986). Thus, a 160-pound (approximately 73-kilogram) person would need 9,125 mg of THC to have a 50% chance of dying. A typical marijuana cigarette weighs one gram and contains roughly 20 mg of THC, suggesting that roughly 450 joints would prove fatal. Furthermore, at least 50% of the THC is destroyed in the burning process or lost to sidestream smoke. Given this loss, 900 joints would be a more appropriate estimate of a fatal amount (Doweiko, 1999). The 900 joints would weigh roughly 2 pounds. Although experienced users tell many exaggerated tales about smoking large amounts of cannabis, this dosage exceeds 100 times the quantity typically consumed by the heaviest users.

Given the limited fear of lethal overdose, marijuana with larger percentages of THC may actually have some benefits. Stronger cannabis may lead to smoking smaller amounts in order to achieve desired effects. Smoking smaller quantities could provide some protection against the health problems normally associated with inhaling smoke. Smokers may take smaller, shorter puffs when using more potent marijuana (Heishman, Stitzer, & Yingling, 1989). Smoking less may decrease the amount of tars and noxious gases inhaled, limiting the risk for mouth, throat, and lung damage (Matthias, Tashkin, Marques-Magallanes, Wilkins, & Simmons, 1997). Obviously, avoiding smoke completely would eliminate these problems. Thus, eating cannabis products may have fewer negative consequences than smoking them. Comparisons between these two ways of administering the drug appear next.

Cannabinoid Administration

A thorough understanding of marijuana's effects requires some knowledge of how it enters a biological system. Drugs can penetrate the body in many ways. Humans inject drugs intramuscularly (into muscles) or intravenously (into veins). They also inject subcutaneously (under the skin), a process known as "skin popping." People snort drugs intranasally. Some substances can be absorbed sublingually by placing them under the tongue. A few drugs can dissolve through the skin in transdermal administration, like the ubiquitous nicotine patch. None of these methods is particularly common for marijuana. Cannabis has two popular routes of administration: inhalation (smoking) and oral ingestion (eating). In addition, researchers have examined intravenous injections of THC and rectal administration via the marijuana suppository. Drug companies have also proposed a deep lung aerosol, a nasal spray, a nasal gel, and a sublingual preparation of synthetic THC (IOM, 1999).

All these alternative techniques for administering the drug remain relatively rare, but inhalation is quite common. In addition to cannabis, humans inhale nicotine, opium, crack and freebase cocaine, methamphetamine, glue, gasoline, and anesthetics. Inhalation serves as one of the fastest modes of administration for any drug, and THC is no exception. Smoke held in the lungs contacts the bloodstream directly through a rich network of capillaries. This blood travels almost immediately to the brain, the site of the majority of cannabinoid receptors. Thus, the first hints of intoxication can appear within 10 seconds of exhaling the smoke (Levinthal, 1999). People may reach their peak blood concentration of THC while still smoking. The rapid absorption during smoking parallels the dramatic increases associated with intravenous doses of THC. THC dissolves readily in fatty tissues of all sorts, but eventually travels through the blood to the liver and kidneys. It is subsequently metabolized and excreted.

Several factors influence the amount of THC absorbed during smoking. Larger puffs held deeply in the lungs for a long time create the most dramatic effects. At least 30% of the THC in cannabis disappears in the combustion of smoking. More cannabinoids escape while the marijuana cigarette burns between puffs (Davis, McDaniel, Cadwell, & Moody, 1984). Some studies suggest that experienced smokers can take in more THC than inexperienced ones, that is, they may inhale more efficiently. Experienced smokers seem to know their lung capacity and understand the amount of smoke that they can hold without coughing. In contrast, inexperienced smokers may take larger puffs that they subsequently cough out, or smaller puffs that do not provide much of the psychoactive chemicals.

In laboratory studies, heavy users absorbed approximately 27% of the THC available in a joint; light users absorbed only about 14% (Ohlsson, et al., 1982; Ohlsson, Agurell, Lindgren, Gillespie, & Hollister, 1985). The increased efficiency of inhalation in experienced users may account for a curious phenomenon called reverse tolerance or sensitization. Many regular users of cannabis report rapid effects at extremely low doses of the drug. Researchers once posited that people grew more and more sensitive to the drug with repeated exposure, allowing them to experience the subjective effects with less and less of the substance.

Limited absorption may contribute to the minimal effects that novice smokers report the first few times they try cannabis. Many eventually learn to inhale and report more impact from the drug. Some never learn to inhale and subsequently run for public office. The amount of THC an individual assimilates while smoking can vary dramatically. Yet users rarely complain about an inability to absorb enough THC. The effects of smoking are rapid, and people can modify their doses quite readily. A detectable increase in dosage is usually a mere puff or two away. Instead, complaints related to smoking concern irritation of the mouth, throat, and lungs. These complaints occasionally lead a smoker to eat marijuana or hashish instead.

Oral administration has the longest history of all the techniques for using drugs, beginning with alcohol consumption around 8000 B.C. (Roueche, 1963). Most substances taken by mouth must travel the entire gastrointestinal tract, which contains several natural barriers to absorption. These barriers are important in helping minimize the toxic effects of many substances. The interior of the stomach, a highly acidic environment, can break down a variety of noxious chemicals. Unfortunately, this environment also neutralizes potentially helpful medications. For example, stomach acid destroys insulin, making oral administration of this drug useless.

Drugs that survive the stomach pass to the small intestine. Membranes between the intestinal wall and surrounding blood capillaries include two layers of fat molecules. Thus, only fat-soluble substances pass into these capillaries, which then send the drug to the liver for further metabolism. Those drugs that survive the liver metabolism reach general circulation. Yet entering the bloodstream is no guarantee of reaching the brain. The blood-brain barrier, a tightly knit bed of capillaries, keeps all but a few of the most lipid-soluble substances from reaching the brain. This barrier separates the brain from the circulatory system, minimizing the impact of any potential toxins.

Given all the natural barriers inherent in oral administration, it remains dramatically slower than inhalation. Eating marijuana may create smaller effects than smoking an equal amount because so much of the drug breaks down in digestion. Eating cannabis or hashish leads to delayed, erratic absorption, depending upon the state of the gut at the time. A gram of cannabis might lead to extreme intoxication when smoked but hardly alter subjective experience if eaten after a full meal.

The concentration of THC also tends to peak much later after oral administration than after smoking. Concentrations peak between 1 and 6 hours after eating THC in chocolate cookies and around 2 hours in sesame oil pills like dronabinol (Ohlsson et al., 1980; Ohlsson et al., 1985; Wall, Sadler, Brine, Harold, & Perez-Reyes, 1983). In one study, marijuana-laced brownies led to some effects in 30 minutes, but peak responses did not occur for 2½ to 3½ hours (Cone, Johnson, Paul, Mell, & Mitchell, 1988). Cannabis resin eaten in a meat sandwich also took at least 2 hours to create peak effects (Law, Mason, Moffat, Gleadle, & King, 1984). Thus, the digestive process decreases the bioavailability of the drug. Bypassing this digestive process via smoking enhances effects. The marijuana suppository also bypasses degradation in the liver and leads to greater availability of THC (Mattes, Shaw, Edling-Owens, Engelman, & ElSohly, 1993).

Despite the potential drawbacks of decreased availability and slower initiation of effects, orally administered marijuana has developed quite a history. Cannabis products heated in oil or butter and combined with sweets have served as confections for centuries (Abel, 1980). Modern recipes for brownies, soup, meatloaf, guacamole, banana bread, and cookies can include cannabis. Hashish recipes for cookies, brownies, and soup are also quite common (Powell, 1971). Tinctures made from marijuana soaked in alcohol also provide a vehicle for oral administration. In addition, dronabinol, the synthetic version of THC dissolved in sesame oil, is marketed in capsules for easy swallowing. Eating the drug avoids the obvious throat and mouth irritation and risk for lung problems that accompany smoking.

THC Metabolism

The amount of time required to metabolize THC has shown considerable variation from person to person and study to study. The period required to eliminate THC from the body should not be confused with the duration of the drug's psychoactive effects. People stop feeling high long before THC has left their bodies. Intoxication rarely last more than a few hours, with orally administered doses lasting longer than smoked cannabis. After an intravenous injection of THC, blood levels peak almost immediately and then decrease by 90% in the first hour. This rapid drop does not mean that the drug has exited the body; it simply leaves the blood to dissolve into fat tissue. THC in the blood partitions into fat tissue, then leaks slowly from fat to be degraded and excreted. Although media accounts of marijuana's effects often treat THC's fat solubility as a novelty, sedatives like the barbiturates and benzodiazapines are stored in fat, too. After the first hour, blood levels of THC do not drop as rapidly. As THC from the blood is eventually excreted in urine and feces, THC stored in fat returns to circulation, but in doses too small to create psychoactive effects.

Researchers express the time required to metabolize a drug as its half-life—the period required to break the dose down to 50% of its original amount. Suppose the half-life of a hypothetical drug was one day. People who absorbed 100 mg of this drug would reduce it to 50 mg in one day. The next day they would again cut the available dose in half, to 25 mg. The next day would decrease the amount to 12.5 mg, and so on. Zeno's paradox would suggest that this consistent splitting in two would actually never lead to a blood level of zero. The amount would decrease by 50% repeatedly, growing smaller and smaller, but it would never disappear. Practically, drugs reach an undetectable concentration after 4.5 or 5 half-lives (Diaz, 1997).

Estimates of the half-life of THC based on urinary excretion show incredible variation. Research estimates of THC's half-life range from as little as 19 hours (Hunt & Jones, 1980) to as much as 4 days (Johansson, Arguell, Hollister, & Halldin, 1988). Early work suggested that users might grow more efficient at metabolizing THC as they gain experience with the drug (Lemberger, Axelrod, & Kopin, 1971). This study found a half-life of 28 hours for chronic smokers, but naive users took 57 hours to metabolize half of the dose. These results had considerable intuitive appeal because they helped explain marijuana tolerance. The authors of this research implied that people grew tolerant to the drug because they essentially eliminated it more readily.

In contrast to evidence of increased rates of metabolism in experienced users, a later study found that people who had received THC each day for two weeks did not metabolize faster than moderate users who had not ingested daily doses (Hunt & Jones, 1980). In addition, the study with the longest estimated half-life used chronic, regular smokers (Johansson et al., 1988). These results seem confusing in light of the previously reported shorter half-life for heavier users. Thus, THC's rate of metabolism may not increase with repeated use. Most studies show a half-life between 1 and 1.5 days (Ohlsson et al., 1982; Ohlsson et al., 1985; Wall et al., 1983). A recent study using an extremely sensitive measuring technique and a two-week follow-up period found THC half-life ranges up to 2.5 days (Huestis & Cone, 1998). This study used the best methods available and suggests that a dose of THC leaves the body completely after 12 or 13 days.

The extreme variation in the estimates for the half-life of THC may stem from studying small samples of people over relatively short durations, using measurement techniques that vary in sensitivity. This large range of estimates likely reflects individual differences among people. Some simply metabolize more quickly than others. Techniques that do not rely on urine samples suggest THC stays in the body even longer. Analysis of fat cells rather than urine samples has revealed that the drug can remain in the body up to a month in some people (Johansson, Noren, Sjovall, & Halldin, 1989). Popular authors imply that this long elimination period is the norm (DuPont, 1984), but many people metabolize THC faster. Despite all the variability in elimination periods, marijuana does appear to have a longer half-life than some other drugs. For example, nicotine's half-life is about 2 hours; caffeine's is 3 to 6 hours (Henningfield, Cohen, & Pickworth, 1993). However, some sedatives that are more fat soluble show half-lives around 2 days or more (Diaz, 1997). Thus, marijuana takes more time to metabolize than some drugs, but less than others.

Popular authors often misinterpret THC's long half-life by frequently implying that intoxication or some sort of residual effect of the drug remains for weeks at a time. Yet intoxication dissipates in a couple of hours. The amount of THC released gradually from fat cells does not create any subjective, cognitive, or emotional effects but may register on drug tests. Thus, a person may test positive for cannabis even a week or two after smoking, when all signs of intoxication have clearly terminated (Zimmer & Morgan, 1997). A number of underground legends suggest that goldenseal, cranberry juice, or various other concoctions might shorten this period of testing positive; no systematic research addresses this question. Drinking enormous quantities of fluids may dilute THC metabolites in the urine and alter the outcome of a test, but these fluids do not actually alter metabolic rate (Coombs & West, 1991).

Cannabinoid Receptors

Once cannabinoids enter the body, they must find a site to create their effects. The quest to understand the biological function of cannabinoids has generated a large body of research. Initial studies tracked radioactive THC through the body. This work revealed that THC attached to all the surfaces of the neuron, suggesting that it might alter the permeability of cell membranes to create its impact (Makriyannis & Rapaka, 1990). Researchers were familiar with the idea of drugs altering membrane permeability because some of alcohol's effects may stem from a comparable process (Doweiko, 1999).

Later work revealed that at least some of marijuana's impact could not arise solely from changes in the permeability of cell membranes. Newer studies found that cannabinoids could inhibit the synthesis of an extremely important compound, cyclic adenosine monophosphate (cyclic AMP or cAMP), which helps initiate nerve impulses (Howlett, Johnson, Melvin, & Milne, 1988). Other drugs that work in this way have special receptors that alter the cAMP. These receptors inhibit adenylyl cyclase (AC), the enzyme used to make cAMP. Examinations of every receptor known to inhibit AC revealed that none of them responded to the cannabinoids.

With all these other receptors ruled out, researchers concluded that cannabis must work via its own site. Investigators soon identified the cannabinoid receptor and mapped its distribution in the brain (Bidaut-Russell, Devane, & Howlett, 1990; Devane, Dysarz, Johnson, Melvin, & Howlett, 1988 Herkenham et al., 1990). The cannabinoid receptors in the nervous system, which are known as the CB1 type, are quite numerous. By way of comparison, CB1 receptors are 10 times more abundant than mu opiod receptors, the sites of action for morphine. After researchers identified the CB1 receptor in the brain, other work revealed a second site of action in the immune system. This receptor was dubbed, surprisingly, CB2.

Contrary to what we all learned in high school courses on biology, receptors are not locks that either open or close in response to some key. The cannabinoid receptors (and all others) are proteins—strings of amino acids that span the membrane of the cell. Some of the amino acids are embedded in the cell membrane; some are implanted inside the cell; others extend outside. Cannabinoids bind with the portion of the receptor outside the cell and trigger activity inside. Different cannabinoids bind in different ways, leading to varied amounts of activity within the nerve cell.

One action triggered by the cannabinoid receptor is the inhibition of AC and subsequent inhibition of cAMP, as mentioned above. Thus, any process that requires cAMP will slow down if the cannabinoid receptor has been activated. The receptor also opens the potassium channels of the neuron, which decreases its rate of firing. Unlike the potassium channels, calcium channels close when the cannabinoid receptor activates. Closed calcium channels decrease the release of neurotransmitters. Thus, by inhibiting cAMP, slowing the nerve's firing rate, and decreasing neurotransmitter release, cannabinoids alter the communication between nerve cells. These actions may account for many of the effects of THC as well as other cannabinoids.

The cannabinoid receptors and their associated brain systems do not work in a vacuum. Any alterations of one neurotransmitter can change the functioning of others. THC clearly creates changes in the dopamine system, as cocaine, amphetamine, nicotine, and alcohol do (Koob & Le Moal, 1997). The cannabinoids can enhance dopamine's activation of movement, suggesting that they might help treat Parkinson's disease (Sanudo-Pena & Walker, 1998). Cannabinoids can inhibit or enhance gamma-aminobutyric acid (GABA), the neurotransmitter that may contribute to alcohol's sedative effects (Pacheco, Ward, & Childers, 1993; Shen, Piser, Seybold, & Thayer, 1996). THC also interferes with acetylcholine, a neurotransmitter involved in memory. The effect on acetylcholine may underlie the memory problems associated with cannabis intoxication. Thus, the cannabinoid receptor and related neurotransmitter systems clearly play an important role in the functioning of the brain.

Human cannabinoid receptors are extremely similar to those found in rodents, suggesting that evidence from animal studies may apply to people (Gerard, Mollereau, Vassart, & Parmentier, 1991). Leeches, mollusks,

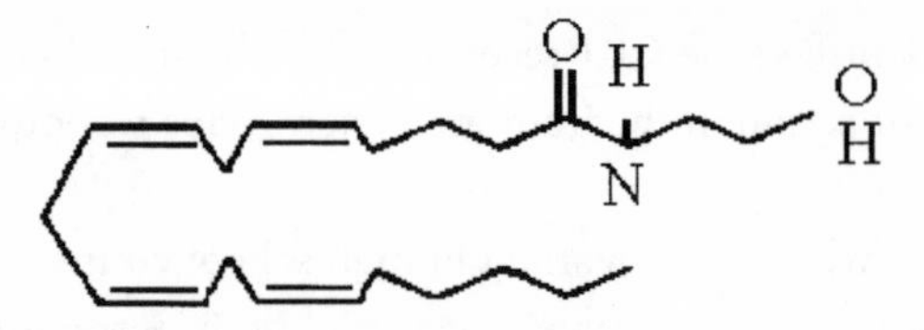

Figure 6.6. Anandamide. This cannabinoid occurs naturally in the body. It is the first endogenous cannabinoid discovered.

chickens, turtles, trout, and fruit flies have cannabinoid receptors, too (Howlett, Evans, & Houston, 1992; Stefano, Salzet, & Salzet, 1997). Even a primitive protozoan, the Hydra, has a cannabinoid receptor that appears to alter its feeding (De Petrocellis, Melck, Bisognor, Milone, & DiMarzo, 1999). The presence of the receptor in such a wide variety of species suggests that it must have an important and universal function. The ubiquitous presence of the CB1 and CB2 receptors inspired a search for substances within the body that might react at these sites.

The Body's Own Cannabinoids

It seems unlikely that so many animals would develop receptors simply to respond to some green weed. The identification of the cannabinoid receptors inspired the search for the body's own substances that might activate them. Studies of the functions of these endogenous cannabinoids could reveal a lot about how marijuana works, as well as how the brain works. Several endogenous chemicals appear to interact with the cannabinoid receptor. The two studied most are arachidonylethanolamine (anandamide) and 2-arachidonolyl-glycerol (2-AG), pictured respectively in figures 6.6 and 6.7.

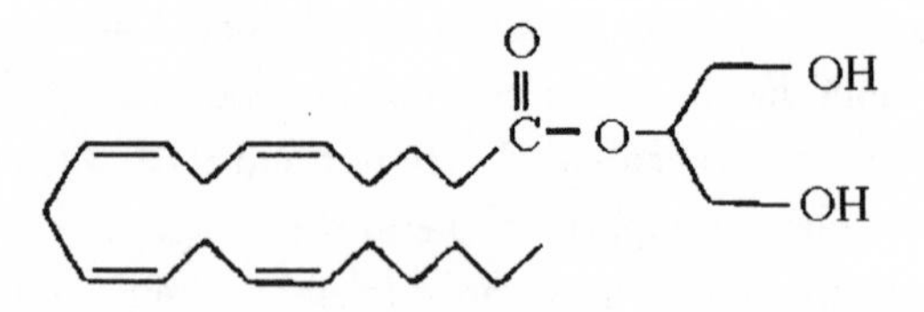

Figure 6.7. 2-arachidonolyl-glycerol (2-AG). This substance is the most abundant endogenous cannabinoid.

The first cannabinoid identified in the body was dubbed anandamide, from the Sanskrit word *ananda*, which means "bliss" or "ecstasy" (Devane et al., 1992). Its actual chemical name is arachidonylethanolamine. As the name suggests, its components include arachidonic acid and ethanolamine. Arachidonic acid serves as the building block of dozens of other chemicals, including some of those involved in the way aspirin works. Ethanolamine is related to alcohol. The anadamide molecule does not particularly resemble THC, but both interact with the cannabinoid receptors. Other natural compounds work on drug receptors but do not share the drug's molecular shape. For example, endorphins, the endogenous opiates, do not look much like opium or morphine. Yet they work at the same sites on the neuron. Clearly, molecules of different shapes may still connect to the same receptor. The critical aspects of the shape of the molecule are not known precisely.

Enzymes metabolize anandamide quite quickly, so its duration of action is shorter than THC's. Anandamide creates less intense effects than THC, perhaps because of this rapid breakdown. It also has only 25 to 50% of THC's affinity for receptors. Still, anandamide creates some of the same reactions. For example, this endogenous cannabinoid induces overeating (Williams & Kirkham, 1999) and lowers activity, body temperature, and pain sensitivity (Fride & Mechoulam, 1993). Researchers find anandamide in many of the brain areas rich in CB1 receptors, including the hippocampus, a structure involved in memory. Thus, this receptor may play a role in the deficits in short-term memory associated with marijuana intoxication. CB1 receptors also appear in the cerebellum, a motor center of the brain. The ability of cannabinoids to relieve spasticity and tremor may involve the receptors in this area (Baker et al., 2000).

Anandamide also appears in the thalamus, a structure involved with pain and emotion. Oddly, the thalamus has relatively few cannabinoid receptors. Perhaps anandamide works on other receptors in this structure. Anandamide works in systems outside the brain as well. It appears in spleen tissue, which has many CB2 receptors, and acts in the immune system (IOM, 1999). Anandamide even inhibits the growth of breast cancer cells (De Petrocellis et al., 1998). Research on anandamide's exact functions in the brain and immune system continues.

Several compounds related to anandamide also bind to cannabinoid receptors, including 2-arachidonolyl-glycerol (2-AG). The compound 2-AG is a prominent constituent of brain tissue, about 170 times more abundant than anandamide (Stella, Schweitzer, & Piomelli, 1997). It clearly interacts with both the CB1 and CB2 receptors (Sugiura et al., 1999; Sugiura et al. 2000). It also alters heart rate and blood pressure in mice (Jarai et al., 2000). Its exact role in other cannabinoid effects is not yet clear. Research on the endogenous cannabinoid system progresses rapidly. Studies have already isolated new substances that bind to cannabinoid receptors, but they have not been identified yet. Their precise biological impact remains unknown.

Conclusions

Marijuana contains more than 400 chemical components; at least 66 of them are cannabinoids unique to the plant. The most prevalent ones include delta-9-THC, cannabinol, and cannabidiol. THC causes cannabis's intoxicating effects. Cannabinol has about one-tenth the psychoactive effects of THC. At high doses, it can increase sleep and decrease body temperature. Cannabinol may decrease THC's psychoactive effects, particularly the stimulating aspects of the subjective experience, but it may also extend the duration of intoxication. Cannabidiol may decrease anxiety and psychotic symptoms as well as minimize seizures. In combination with THC, cannabidiol may increase THC concentrations, slow its metabolism, and limit any anxious or paranoid feelings associated with intoxication. Researchers have identified dozens of other cannabinoids, many with shapes similar to these three. All are soluble in fat, but their other chemical properties are not understood completely.

Many different preparations of cannabis exist, including wide varieties of hashish, hash oil, and marijuana. Their potencies can vary dramatically, with hashish containing up to 50% THC and hash oil running as high as 70% THC. Cannabis itself is often between 2 and 4% THC, with some claims of potency reaching markedly higher. Concerns about dramatic increases in potency over the last 30 years may stem from poorly analyzed or misrepresentative samples of cannabis from the 1970s. Assertions about increases in potency of 10 to 100 times seem extremely unlikely. THC concentrations have probably increased by a factor of 2 or 3. These increases may not justify alarm. THC is not toxic at high doses like alcohol, nicotine, or many other common drugs. High-potency marijuana may actually minimize risk for lung problems because less is required to achieve desired effects.

Users often smoke cannabis products and occasionally eat them. Researchers are currently experimenting with other ways to administer THC, including a nasal spray and rectal suppository. Smoke inhalation provides a rapid absorption of THC into the blood and brain, creating striking changes in subjective experience that often last a couple of hours. Ingesting marijuana or hashish requires digestion in the gastrointestinal tract and liver, delaying and reducing effects but lengthening their duration. Although intoxication rarely lasts more than a few hours, the complete elimination of THC clearly takes at least a few days. The chemical can remain in fat cells for up to one month. The fat cells eventually release the THC back into the blood stream, but in quantities too small to have any subjective effect. The liver breaks down this released THC and its metabolites are excreted.

In an effort to understand which neurotransmitter systems create THC's effects, investigators eventually identified two receptors that respond specifically to the cannabinoids. One receptor (CB1) exists in the brain and appears in high concentrations in areas involved with memory and motor control. The other receptor (CB2) is most prevalent in the immune system. The identification of this new neurotransmitter system has generated considerable research and reveals that the brain remains more complex than previously thought. The presence of these receptors inspired a search for the body's own chemicals that may activate them. Research has uncovered two natural cannabinoids; anandamide and 2-AG. These appear to mimic some of THC's effects, though they are less potent and have a shorter duration of action. Future research in this area will likely continue to inform us about the way this drug works as well as how the brain functions.

7
Marijuana's Health Effects

Concerns about marijuana's potential impact on health have generated volumes of research, numerous conferences, and considerable controversy. This chapter addresses marijuana's toxicity, as well as its impact on mental illness, the brain, the pulmonary system, reproduction, pregnancy, and immune function. In general, the drug is incapable of creating an overdose. It can exacerbate the symptoms of some mental disorders but does not appear to cause them. Data fail to show any marijuana-induced changes in brain structure, but long-term exposure to the drug alters the way the brain functions during complex tasks. People who smoke cannabis but not cigarettes have yet to show severe pulmonary problems like lung cancer or emphysema, but milder respiratory problems do appear. Large doses of cannabinoids can cause temporary changes in reproductive hormones and sperm, but these effects reverse with abstinence. The role of cannabinoids in immune function appears extremely complex, but data have yet to show that smoking marijuana increases the rates of infectious disease in humans. Details on each of these areas appear below.

Toxicity

Media reports highlight tales of increased THC content in new strains of cannabis, leading some to worry about the potential of a fatal overdose. Yet cannabis is essentially nontoxic. No one has ever died of THC poisoning (Iversen, 2000). Extrapolations from animal research suggest a lethal dose of THC would require 125 mg of the drug per kilogram of body weight (Nahas, 1986). A 160-pound (73-kilogram) person would require 9,125 mg of THC to receive a fatal dose. Most marijuana cigarettes weigh one gram and contain 20 mg of THC. Thus, this 160-pound person would require all the THC in over 450 joints to reach a lethal dose. At least 50% of THC, however, is lost to sidestream smoke. Therefore, a lethal dose would actually require smoking 900 joints. If some new strain of genetically engineered super cannabis contained 10 times the usual amount of THC, a lethal dose would require 90 joints. If a user could finish one of these high-powered cannabis cigarettes in 10 minutes, he or she would have to smoke for 15 consecutive hours to reach a lethal quantity. Thus, even the most devoted pothead with marijuana of legendary strength could not stumble upon a fatal overdose. In contrast, alcohol and aspirin poison thousands of people each year (Doweiko, 1999).

Mental Illness

Concerns about marijuana leading to psychological problems are at least 100 years old. When India was a colony, the British government sponsored the Indian Hemp Drugs Commission, which investigated claims that cannabis somehow caused an increase in the number of patients in mental institutions. The commission found that marijuana did not cause mental illness (Indian Hemp Drugs Commission [IHDC], 1894). These results were largely ignored. Propaganda from the United States in the 1930s implied that smoking cannabis invariably led to insanity (e.g., Fossier, 1931; Rowell & Rowell, 1939). Modern authors also suggest that the drug creates mental illnesses, including panic and psychosis (Gorman, 1996; Lapey, 1996). A close look at the data reveals that many people with psychological problems smoke marijuana, but it does not cause their disorders. Yet some people with mental illness may find that the drug aggravates their symptoms. Descriptions of the studies related to this topic follow.

Anxiety Disorders

Marijuana intoxication can heighten anxiety, but this brief reaction should not be confused with an anxiety disorder. Anxious reactions are rare but appear most often in inexperienced users and after eating rather than smoking cannabis. Inexperienced users are more likely to panic, particularly if they smoke a great deal in a short time. Eating hashish can also lead to feeling fearful or alarmed, particularly at high doses. Orally administered THC may take a couple of hours to create subjective effects. People who are unaware of this lag often think that their initial dose is an insufficient amount to alter consciousness. They eat a sufficient amount but fail to wait long enough for the drug to take effect. Thinking that they have not consumed enough, they eat more, and later find themselves markedly more intoxicated than they had planned. These inadvertent overdoses can create many signs of discomfort, including anxiety, paranoia, and visual hallucinations.

Many experienced users know how to titrate smoked doses to avoid any adverse emotional responses. They frequently approach cannabis of unknown strength with caution, smoking a small amount and waiting for its effects to peak before using more. In addition, adverse reactions to the drug are temporary. People experiencing marijuana-induced anxiety usually respond well to simple reassurance. Experienced users find that the drug actually decreases anxiety (Grinspoon & Bakalar, 1997). A recent longitudinal study of more than 800 New Zealanders found no link between cannabis and either anxiety disorders or depression (McGee, Williams, Poulton, & Moffitt, 2000). Thus, marijuana seems an unlikely cause of any clinical anxiety or depressive disorder. Nevertheless, people who are prone to panic or who find changes in consciousness disturbing should avoid cannabis and other psychoactive drugs.

Psychotic Disorders

Cannabinoid intoxication can also mimic certain aspects of psychoses like schizophrenia. These psychotic disorders typically include odd thoughts, auditory hallucinations, and inappropriate emotions. An odd thought typical of psychosis must be completely implausible within the person's culture. For example, a psychotic individual might have the odd thought that other people are inserting ideas into his head. Auditory hallucinations usually include hearing voices that do not exist. Inappropriate emotions might include smiling when frightened or sad. Large doses of eaten marijuana or hashish can create comparable symptoms, but this cannabis psychosis is not the same as schizophrenia. It usually lacks the formal thought problems and inappropriate emotions (Basu, Malhotra, Bhagat,

& Varma, 1999). It also dissipates relatively quickly, while schizophrenia remains a chronic mental illness. Other drugs, particularly the hallucinogens, create these symptoms as well. LSD, mescaline, and psychedelic mushrooms lead to erratic thoughts and strange speech patterns typically found in psychotic individuals. Extended use of cocaine or amphetamine can also lead to the paranoid, irrational behavior common to some forms of schizophrenia.

These links between marijuana and odd symptoms reveal valuable information about the way the brain works and suggest that the cannabinoid system may be involved in psychotic disorders. Yet they do not support the idea that cannabis causes mental illness. People with schizophrenia probably will experience fewer problems if they stay away from all recreational substances, but the drugs do not cause their disorder. For example, one study of the Swedish military attempted to predict psychotic breaks from marijuana consumption. Users were more than twice as likely to develop schizophrenia than nonusers. Heavy users (those who had consumed the drug 50 times or more) were 6 times as likely to develop schizophrenia. Yet those who smoked marijuana used a variety of other drugs, which could have easily exacerbated symptoms, too. This study also did not note if people had experienced symptoms prior to their drug use. Perhaps people had odd symptoms but did not experience a psychotic break for years. In that time they may have happened to smoke marijuana. They may have even sought the drug because of their initial odd symptoms (Andreasson, Allebeck, & Rydberg, 1989).

Other work suggests that symptoms precede rather than follow marijuana consumption (Thornicroft, 1990). Perhaps psychotic individuals are more likely to experiment with cannabis. Another argument against the idea that marijuana causes schizophrenia concerns the prevalence of cannabis use and psychotic disorders. If marijuana caused schizophrenia, rates of the disorder should increase with the prevalence of the use of the drug. In fact, schizophrenia is no more prevalent during historical periods of extensive cannabis consumption than at any other time (Hall & Solowij, 1998). These findings suggest marijuana probably does not cause schizophrenia.

Although the drug does not cause the disorder, marijuana may exacerbate psychotic symptoms in schizophrenics. A few studies have found that people with schizophrenia use cannabis frequently, often prior to psychotic episodes (Linzen, Dingemans, & Lenior, 1994; Thornicroft, 1990). Schizophrenics also report preferring cannabis to other illicit drugs

(Dixon, Haas, Weiden, Sweeney, & Frances, 1991). Psychotics who use cannabis regularly have more hospitalizations and worse symptoms (Caspari, 1999). Some authors suggest that the disorder may involve the endogenous cannabinoid system. Evidence for this connection comes from spinal taps that reveal higher levels of endogenous cannabinoids in schizophrenics. Thus, psychotic problems may stem from dysfunctions in this neurotransmitter system (Leweke, Giuffrida, Wurster, Emrich, & Piomelli, 1999). Further work on this topic can reveal a great deal about the role of cannabinoids in brain function. The cannabinoid system obviously plays an important part in our experience of consciousness. If this system goes awry, people may show psychotic symptoms.

Antisocial Behavior

Other data confirm the idea that mental illness may contribute to cannabis consumption rather than marijuana creating mental illness. Marijuana may not cause mental illness, but people with mental illness may use a lot of marijuana. A commendable longitudinal study of more than 800 New Zealanders found that those who had mental health problems at age 15 were more likely to smoke marijuana at age 18. This finding might suggest some form of self-medication. People experiencing symptoms might turn to marijuana for relief. The disorders most likely to predict later cannabis use were those most associated with breaking rules in general—conduct disorder and oppositional-defiant disorder. These two childhood problems stereotypically include a disregard for regulations and authority. This finding suggests that adolescents with little respect for the law are more likely to break it by smoking cannabis.

Although mental illness at age 15 led to more marijuana use at age 18, marijuana use at age 15 did not lead to mental health problems at age 18. At least for these ages, mental health had a bigger effect on cannabis consumption than cannabis consumption had on mental health. As participants grew older, the links between mental illness and cannabis changed. Although mental illness at age 15 predicted marijuana use at age 18, mental illness at age 18 did not predict cannabis use at age 21. Perhaps mental illness only increases cannabis use at a specific stage of development.

One finding from this study initially appears consistent with the idea that marijuana increased mental illness. For men only, smoking marijuana at age 18 predicted specific kinds of mental health problems at age 21.

Smoking marijuana did not lead to anxiety or depression. Instead, men who smoked marijuana at age 18 were more likely to show cannabis dependence, alcohol problems, and antisocial personality disorder at age 21 (McGee et al., 2000). These data must be interpreted with caution. The idea that drug use at one age predicts drug problems at a later age is not new. Longer use of cannabis may increase the chance of problems with that drug or alcohol. These findings may not fit most people's idea of marijuana causing insanity, but the more cannabis a young man consumes, the more likely he is to develop dependence on it or alcohol.

The results concerning antisocial personality disorder may be a bit more complex. The study reports that smoking cannabis at age 18 predicted antisocial personality disorder at age 21. It is important to understand the diagnosis of antisocial personality in order to interpret this finding. Personality disorders refer to enduring, long-term characteristics that may last a lifetime. Yet antisocial personality disorder cannot be diagnosed before age 18. This diagnosis developed from definitions of psychopathy and sociopathy and typically involves a lifelong history of aggression, crime, and lying. Antisocial personality disorder is very much like an adult version of conduct disorder. In fact, to receive the diagnosis of antisocial personality, one must have had conduct disorder during childhood. Some men in this study may have had the traits of this personality disorder all their lives.

Conduct disorder is one of the problems that predicted marijuana consumption when the men were younger. They could not receive the diagnosis of antisocial personality until they were old enough, but they were already destroying property, getting in fights, and using drugs. Thus, cannabis use was probably a simple part of their general tendency to disregard rules. Although marijuana use may precede the diagnosis of antisocial personality, this situation may arise because the disorder cannot be diagnosed until age 18. The marijuana use actually may serve as more of a correlate of underlying deviance and disrespect for the law. Cannabis may not cause psychopathy, but psychopaths often smoke cannabis.

The Brain

Structural Damage

Careful research on humans shows no structural changes associated with chronic cannabis exposure in adulthood. One of the first studies to address this question used a controversial technique (air encephalography) to measure the size of brain ventricles in 10 young men who complained of neurological symptoms after smoking cannabis regularly for up to 11 years. This study revealed larger ventricles in these users relative to a control group, suggesting that their brains had atrophied (Campbell, Evans, Thomson, & Williams, 1971). The results received widespread media attention. Critics of the study emphasized that the cannabis users had also consumed many other drugs. Critics also explained that the technique employed does not provide accurate assessments of brain volume. Subsequent work using improved measurements failed to replicate these findings, suggesting no cerebral atrophy in chronic cannabis users (Stefanis, 1976). These new data, however, received little media attention.

Computer assisted tomography (CAT) scans provided more accurate measurements of brain structure, with none of the controversy that accompanied air encephalography. Using this improved technique, a study of a dozen people who had consumed 5 joints per day for 5 years showed no evidence of cerebral atrophy (Co, Goodwin, Gado, Mikhael, & Hill, 1977). Another study of 19 people who consumed 25 to 62 joints per month for at least a year found no irregularities in their CAT scans (Kuehnle, Mendelson, & David, 1977).

A third project examined CAT scans in a dozen users who had smoked at least a gram of cannabis per day for between 6 and 20 years. Again, researchers found no evidence of structural damage or cerebral atrophy (Hannerz & Hindmarsh, 1983). A recent study using magnetic resonance imaging (MRI), a relatively new technique with improved accuracy for measuring tissue volume, unexpectedly found larger brain ventricles in nonusers—as if the marijuana had somehow protected the brains of smokers from atrophy. This study found no abnormalities in the brains of users, even though they smoked cannabis twice a day for at least 2 years (Block, O'Leary, Ehrhardt, et al., 2000). Despite these findings, no one suggested that marijuana protected people from cerebral atrophy.

In contrast to all of these studies that found no structural changes in adults, adolescent users of marijuana may alter the development of their brains. In a new study using MRI, researchers reported smaller brains, a lower percentage of gray matter, and a higher percentage of white matter in adults who started smoking marijuana before age 17 (Wilson et al., 2000). This result is particularly alarming because it may indicate interrupted brain development rather than atrophy. THC may have impaired the natural changes in brain structure that accompany adolescence. Some of these structural differences are consistent with the idea that THC's impact on certain hormones may interfere with natural changes in the brain.

The participants who started smoking earlier also had smaller bodies, which is consistent with arrested development. The men who started smoking before age 17 weighed an average of 20 pounds less and were an average of 3 inches shorter. The women weighed 7 pounds less but were equal in height. The differences in brain structure did not correlate with the number of years of use. This result suggests that cannabis exposure during a critical period might interfere with brain development, but further exposure after that period may have little additional impact.

This study does not offer conclusive proof that marijuana created these structural changes in the brains of young users. Participants were not randomly assigned to smoke cannabis before age 17. A subset of individuals with smaller brains may have chosen to use the drug. More important, those who started using marijuana earlier also used more of other illicit drugs. Some other drug may have had this impact on brain structure. In addition, the sample was relatively small (57 people). A replication with a large sample of users who did not differ on other drug consumption could prove very helpful. Despite these caveats, the potential for interference in brain development offers more support for the National Organization for the Reform of Marijuana Law's recommendation that cannabis use not begin before adulthood (NORML, 1996a).

Brain Function

Although advanced techniques in brain imaging reveal that those who start cannabis use in adulthood show no marijuana-induced changes in brain structure, chronic cannabis consumption might alter brain function. New measures of function include brain imaging techniques that measure blood flow and metabolism. Other approaches use the electroencephalogram to assess brain waves, which relate to the way that people process information. These measures are more sensitive to changes in the brain that do not involve flagrant alteration in anatomy. They also reveal marijuana-related changes that did not appear in MRIs or CAT scans.

Cannabis researchers have examined cerebral blood flow (CBF) and brain metabolism, particularly in the cerebellum. This brain structure contains many cannabinoid receptors. It also plays an important role in the perception of time, which usually goes awry during cannabis intoxication. Thus, it seemed a prime target for altered function in chronic cannabis users. The first study of this type revealed reduced blood flow throughout the brain for 9 chronic, heavy smokers who had used cannabis for 10 years. These results were confounded, however, because some participants took benzodiazepines before the assessment procedure. Benzodiazepines clearly limit cerebral blood flow, which may account for these results (Solowij, 1998; Tunving, Thulin, Risberg, & Warkentin, 1986).

A different study, which avoided any problem with benzodiazepines, revealed that experienced users had lower CBF than controls (Mathew & Wilson, 1992). Other researchers studied 17 people who smoked an average of 17 times per week for at least 2 years and found decreased CBF in the cerebellum (Block, O'Leary, Hichwa, et al., 2000). Another study using positron emission tomography (PET), a nuclear imaging technique, revealed lower metabolic activity in the cerebellum for 8 chronic users who smoked at least weekly for over 5 years (Volkow et al., 1996). Although not all studies of this type reveal differences, the sum of these data suggest that frequent, chronic consumption of cannabis has a detectable impact on the functioning of the cerebellum, lowering blood flow and metabolism in the area.

The meaning of the decreases in CBF and metabolism is not particularly clear. During marijuana intoxication, people usually show increased blood flow to the cerebellum. Perhaps chronic exposure to THC leads the brain to adapt. Heightened cerebellar blood flow may lead the brain to decrease the number of cannabinoid receptors. Research on animals is consistent with this idea. This decrease or down-regulation in receptors may then lead to reduced blood flow to the cerebellum. Chronic smokers can then normalize the blood flow to the cerebellum by using cannabis, and the process might repeat itself. Experienced users might need to smoke simply to normalize the blood flow to the cerebellum. Further work might examine if reduced cerebral blood flow serves as the source of continued cannabis consumption. New research with functional MRI, which can examine the workings of the brain during cognitive tasks, might reveal reductions of cerebellar blood flow during craving for cannabis, followed by increases in blood flow during administration.

Other studies of cannabis and brain function focus on the electroencephalograph (EEG), an indicator of brain waves. Chronic exposure to large doses of cannabis clearly alters the EEG in animals (Solowij, 1998).

Initial studies in chronic human users showed no abnormalities as long as participants were not intoxicated (Rubin & Comitas, 1975; Karacan et al., 1976; Stefanis, 1976). Nevertheless, critics of this work emphasize that these studies focused on the most functional users and failed to use modern analytic techniques. Other studies reveal deviant EEGs in chronic users, but results do not appear consistently. Some studies report more of certain types of brain waves in chronic users. Others report less of the same waves. Still others show no differences at all (Solowij, 1998). Details of these studies follow.

The most consistent effect of chronic marijuana consumption on EEG involves alterations in alpha waves in the frontal cortex. Alpha waves cycle at 7.5 to 12.5 hertz and usually indicate a state of quiet relaxation. People with an average of 10 years of daily marijuana use show more power in these alpha waves. Their frontal alpha waves also show greater coherence, meaning that the left and right sides of their brains seem to emit these waves at the same time. The meaning of this greater coherence in brain waves remains unclear. Greater brain wave coherence may mean a state of relaxation, sedation, or inattentiveness. Participants in this study had refrained from using cannabis for 24 hours, so it is unlikely that this effect stems from intoxication. These increases in frontal alpha waves parallel the effects of acute THC intoxication (Struve et al., 1999). Several studies reveal that alpha waves increase after THC ingestion, particularly during periods of subjective euphoria (e.g., Lukas, Mendelson, & Benedikt, 1995). Perhaps chronic users eventually feel intoxicated, even in the absence of the drug, and show brain waves that parallel the intoxicated state even during sobriety.

Chronic cannabis users have also shown deviant EEGs in response to certain difficult tasks. Lights, tones, and other events can elicit changes in brain waves when people attend to them. These event-related potentials reveal that chronic users do not process information as efficiently as nonusers or ex-users. They also may be more distractible. For example, in a series of studies, participants tried to distinguish among very similar tones. They were instructed to press a button in response to one type of tone (the target), but do nothing in response to the others. People who smoked 2 to 7 times per week for an average of 10 years performed more poorly on this task. They failed to press the button on some of the targets and accidentally pressed it in response to some of the nontarget tones. They also showed deviant brain waves in response to the tones.

In this study, the brain waves of chronic users suggested that they had difficulty discriminating between tones. Nonusers showed large brain wave changes in reaction to the target tones, but no changes in response to the other tones. Compared to the nonusers, chronic users showed less of a change in reaction to the targets and more of a change in reaction to the irrelevant tones. These results suggest that chronic users have trouble separating the relevant from the irrelevant; they do not process information as efficiently. Ex-users who averaged 2 years of abstinence showed some recovery in their brain waves but still had deviant event-related potentials when compared to nonusers.

Another type of event-related potential that appears to vary with chronic cannabis consumption is called the P50. The P50 is a positive brain wave that appears about 50 milliseconds after people hear a clicking sound. When two clicks are presented quickly, one right after the other, an interesting phenomenon occurs. The brain generates a normal P50 wave to the first click, but because the person is still processing the first click, the P50 to the second click is smaller than usual. It's as if the brain filters out some aspect of the second click while it processes the first one. The smaller response to the second click is sometimes referred to as gating, as if the brain closes a gate while processing the first click so that new information from the second click does not interfere.

Some people are better at gating than others. People with poor gating often report many intrusive thoughts, as if too much information reaches their brain at one time. For example, people with schizophrenia often report hearing voices or having racing, tangential thoughts. They also show reduced P50 gating. Yet the medications that alleviate their symptoms can also improve their P50 gating (Light, Geyer, Clementz, Cadenhead, & Braff, 2000). Combat veterans suffering from post-traumatic stress disorder, a condition that typically includes intrusive thoughts with aversive content, also show reduced P50 gating (Neylan et al., 1997). The exact meaning of this event-related response is unknown, but it suggests problems in processing information. People who show this type of information processing deficit may have problems filtering out extraneous information when they try to concentrate.

Chronic users of cannabis also have reduced gating (Patrick et al., 1999). In this study, users had smoked an average of 13 joints per week for an average of 13.5 years. The researchers carefully interviewed these users to ensure that they did not have a psychiatric disorder that might have created deviant brain waves. The users showed significantly less P50 gating than the nonusers. These people may have had deviant gating prior to their use of marijuana, but the data are consistent with the idea that cannabis altered their brain function. Only a longitudinal study that randomly assigns people to smoke marijuana can reveal if the drug is the actual source of the different information processing. Nevertheless, these data suggest that chronic marijuana smokers might be more distractible or have more intrusive thoughts. A replication of this study and research on the subjective experience of consciousness in chronic users might reveal more about this phenomenon.

These data from studies of cerebral blood flow, EEG, and event-related potentials all suggest that information processing and brain functions alter after chronic consumption of cannabis. Some functions may recover after extended periods of abstinence. The practical implications of these changes remain unclear. Subtle deviations in brain function have not translated into deficits on meaningful tasks in daily life. Nevertheless, this evidence supports the idea that the drug alters the way people process information. These measures may be markedly more sensitive than cognitive and neuropsychological tests that reveal few cannabis-induced deficits. Subtle changes in cognitive processing may precede more dramatic problems in brain function, providing a warning to heavy users to stop before obvious cognitive deficits might develop.

The Pulmonary System

Overview

People who smoke cannabis but not cigarettes rarely experience lung problems. Yet the potential for marijuana-induced pulmonary troubles remains. Conclusive proof of marijuana's negative impact on the lungs of humans will require decades of research. Inhaled particles, gases, and heat take time to create disease. Comparable challenges arose in the quest to prove that cigarettes caused lung problems, a task that required nearly 40 years of work. More people die from smoking tobacco than any other single cause. Tobacco smoke and cannabis smoke are very comparable, suggesting that the potential for both to contribute to lung disease is very high (Iversen, 2000). Widespread marijuana smoking did not appear in some countries until fairly recently. Many chronic cannabis smokers are still too young to experience severe pulmonary problems. Nevertheless, chronic users of cannabis do show adverse respiratory symptoms, including cough, phlegm, wheezing, and bronchitis. They also show changes in their bronchial cells that parallel those seen in the early stages of lung cancer (Zimmer & Morgan, 1997).

Respiratory Illness

Current studies have found little evidence of marijuana-induced increases in respiratory illnesses. A review of a large sample of hospital records revealed that 36% of daily marijuana smokers saw a physician for colds, flu, or bronchitis in a six-year period. Only slightly fewer (33%) of the nonsmokers sought treatment for these same problems (Polen, 1993). These data suggest that cannabis consumption does not create meaningful increases in the rates of respiratory illnesses. Nevertheless, other work reveals more symptoms of bronchitis, including chronic cough and phlegm production, in heavy marijuana smokers. A study of daily cannabis users who did not smoke cigarettes showed that they had a higher rate of these symptoms than nonsmokers. Tobacco smokers and people who smoked both substances showed more of these symptoms, too (Tashkin et al., 1987). Another study confirmed these results but found that people who smoked both tobacco and marijuana were more likely to develop bronchitis than those who smoked only one or the other (Bloom, Kaltenborn, Paoletti, Camilli, & Leibowitz, 1987). Thus, chronic, heavy use of cannabis can create respiratory problems comparable to those that tobacco creates.

Lung Function

Research concerning marijuana's impact on lung function has produced mixed results. One study of chronic obstructive pulmonary disease (COPD), a disorder of the lung airways, found no difference between marijuana smokers and nonsmokers. This work relied on a measure of the volume of air that people can expel from their lungs in one second. People who can force more air from their lungs have fewer obstructions in their respiratory tracts. Consumers of tobacco cigarettes invariably show more and more obstructions each year that they smoke, suggesting blocked airways. Yet people who had smoked 2 to 3 marijuana joints per day for 15 years did not differ significantly from others who did not smoke at all (Tashkin, Simmons, Sherrill, & Coulson, 1997). These results suggest that cannabis use may not lead to emphysema.

In contrast, a study of people who smoked an average of one joint per day revealed significant impairment in lung function (Bloom et al., 1987). There were no obvious differences between these two samples to account for the different effects. Thus, marijuana may or may not meaningfully impair the functions of the lungs. Further work can help illuminate the impact of the drug on chronic obstructive pulmonary disease.

Lung Airway Problems

A number of problems with the airways of the lung can appear during examinations with a bronchoscope. This assessment technique can reveal damage that occurs prior to obvious deficits in lung function. Visual inspections of the lungs revealed that people who smoked 5 joints a week for 2 years had more redness, swelling, and mucous. People who smoked both cannabis and tobacco had particularly bad symptoms (Roth et al., 1998). Thus, even without creating emphysema, marijuana can alter the bronchial tract.

Biopsies taken from some of these people revealed that marijuana smokers had more abnormal cells in their lungs. For example, many lung cells normally have cilia, small hairs that help clear the lungs of particles. In cannabis smokers, many of these ciliated cells had transformed into cells more similar to skin. The changes were particularly common among people who smoked both cannabis and tobacco. These sorts of cellular transformations are particularly alarming because they may be the beginnings of the development of lung cancer.

Cancer

Currently, no data reveal definitive increases in rates of lung cancer among people who smoke marijuana but not tobacco. A retrospective study of over 64,000 patients showed no increases in risk for many types of cancer once alcohol and cigarette use were controlled (Sidney, Quesenberry, Friedman, & Tekawa, 1997). Nevertheless, a few lines of research suggest that cases of cannabis-induced lung cancer may appear in the years ahead. THC is not carcinogenic itself. Yet when isolated cells are exposed to marijuana smoke, they change in ways that parallel the early stages of cancer (Leuchtenberger, 1983). Biopsies taken from the lung tissue of cannabis users reveal cellular changes that could lead to tumors (Roth et al., 1996). A number of reports suggest considerable marijuana use among young people with cancers of the lung, oral cavity, and esophagus (IOM, 1999). These data are comparable to early studies of tobacco and cancer and suggest that cannabis smoke is capable of damaging the bronchial system in ways that may lead to tumors.

Pulmonary Harm Reduction

Drug lore suggests that certain strategies may minimize marijuana's potential harm to the lungs. These include ingesting the drug orally, using water pipes or vaporizers, refraining from holding smoke down in the lungs for extended periods, and smoking stronger cannabis. Research suggests that results for some of these strategies are mixed. Obviously, eating marijuana or hashish will have no impact on the respiratory system. Water pipes that cool the smoke will decrease the negative effects of heat. Yet despite popular belief, water pipes do not appear to decrease the amount of tar and particles in smoke (Doblin, 1994). In addition, these pipes may filter out some of the THC, leading users to smoke more cannabis than they might without a pipe. Smoking more may create increased deposits of tar and particles in the lungs. Thus, the water pipe is not a panacea for all cannabis-induced respiratory problems. Its cooling properties may help limit lung damage caused by heat, but other effects are limited.

Another gadget designed to lower exposure to carcinogens is the vaporizer. This pipe uses a hot plate to heat marijuana to the point where cannabinoids vaporize. This temperature should be below the level where carcinogenic hydrocarbons burn. Users can then inhale the vapor, which ideally would contain more THC with fewer contaminants and less tar. Counter to the intentions of the inventors of this machine, it creates a vapor with an unusually low amount of the psychoactive THC and a high amount of the less active cannabinol. An unfiltered joint actually provides a better ratio of THC to tar. Although the benefits of cooler smoke remain, the vaporizer is not the ideal preventer of marijuana-induced lung problems either (Gieringer, 1996).

An additional strategy for reducing lung damage associated with smoking concerns the length of time users keep smoke inside their bodies. The common habit of holding smoke in the lungs for long periods likely increases tar deposits, which undoubtedly add to respiratory problems. Although many experienced users swear by this habit, two studies show that holding "hits" longer does not appear to lead to greater changes in mood (Zacny & Chait, 1989, 1991). Holding one's breath without inhaling any smoke at all can certainly lead to a light-headed, dizzy experience of consciousness easily confused with marijuana's effects. For example, one study revealed that holding one's breath for a long time altered cognitive abilities even when smoking a placebo (Block, Farinpour, & Braverman, 1992). Thus, users who are committed to holding their breath might find that exhaling cannabis smoke beforehand could lead to comparable changes in mood with markedly less risk of lung damage. Otherwise, exhaling soon after inhaling should produce identical subjective experiences with markedly less potential for respiratory injury.

A final approach to preventing harm to the lungs concerns smoking potent marijuana. Cannabis with higher amounts of THC can provide the same subjective experience with a reduced intake of smoke and accompanying deposits of tar. One study showed that 10 regular users deposited less tar when smoking marijuana that was approximately 4% THC than they did while smoking marijuana that was approximately 2% THC. These results suggest that stronger cannabis may actually decrease the risk of respiratory problems (Matthias et al., 1997). British police have confiscated marijuana with THC concentrations around 20% (House of Lords, 1998). Perhaps cannabis this strong could create subjective effects with little exposure to damaging tars. The potential to smoke less while maintaining the same subjective state will undoubtedly help decrease lung and respiratory troubles.

Reproduction

Several interesting lines of research suggest that the cannabinoids play an important role in the function of sex hormones and sperm. This work has led to growing concerns about marijuana's impact on fertility and birth defects. Unfortunately, some enthusiastic efforts to prevent drug use have overstated the data, implying that consumption can lead to permanent infertility or drastic birth defects. (My junior high school health teacher suggested that the children of marijuana smokers would be born with one eye in the center of their foreheads.) Extremely large doses of cannabis could, in theory, decrease fertility in humans. Nevertheless, research has yet to show a definitive decrease in reproductive function in people who smoke marijuana. The drug also has not been linked to human birth defects.

Studies of primates show that THC can alter sex hormones. Large doses of the drug can decrease hormones central to menstruation, including follicle-stimulating hormone, luteinizing hormone, and progesterone. These effects require an injected dose of 2.5 mg of THC per kilogram of body weight, a dose comparable to a 130-pound woman smoking 7 joints per day. Monkeys who received this dose for 18 consecutive days did not ovulate. Nevertheless, with chronic administration of the drug for up to a year, monkeys developed tolerance to this effect and began ovulating again. Comparable results appear for rats and rabbits. Thus, large doses can disrupt female sex hormones, but functions return to normal with time (Smith, Almirez, Scher, & Asch, 1984).

Research on marijuana's impact on fertility in human women offers mixed results. One study showed increased rates of marijuana use in women who are infertile. In the infertile group, 61% of the women had used marijuana. In the fertile group, only 53% had smoked cannabis. Nevertheless, infertile women did not use marijuana more often or for a longer period than those who were fertile. The use of cocaine had a much larger impact on fertility. Once the investigators controlled for cocaine consumption and other factors that contribute to infertility, the impact of cannabis decreased (Mueller, Daling, Weiss, & Moore, 1990). In contrast, a later study found that women who used marijuana regularly conceived more quickly than women who did not use the drug (Joesof, Beral, Aral, Rolfs, & Cramer, 1993). Perhaps these data say more about marijuana's alleged functioning as an aphrodisiac than anything about a direct impact on fertility.

Animal and human research also reveals that THC may interfere with the production of sperm. Extensive studies of sea urchin sperm reveal that cannabinoids decrease their capacity to fertilize an egg (Schuel et al., 1999). In addition, THC lowers the motility of bull sperm in a petri dish (Shahar & Bino, 1974). THC treatments of 5 mg per kilogram of body weight (18 joints per day for a 160-pound man) created twice as many abnormal sperm as usual in mice (Zimmerman, Zimmerman, & Raj, 1979). Anyone intending to breed mice, cows, or sea urchins should keep them away from large doses of cannabinoids. In human research, men who smoked an average of 8 joints per day for one month showed significant drops in sperm count and motility (Hembree, Nahas, Zeidenberg, & Huang, 1979). Nevertheless, their sperm count and functions did not fall to abnormal levels and returned to normal after the study ended. Despite all this work, no studies show an actual decrease in fertility for men who use cannabis.

Pregnancy

The behavior of pregnant women has become a giant public health concern that may mirror attitudes about sex and sexism (Stoltenberg, 1988). Some states may prosecute pregnant women for child abuse if they test positive for drugs. These statutes may lead pregnant women to avoid prenatal health care, dramatically increasing the chances of problematic deliveries. Laws like this assume that illicit drugs damage the fetus. Research designed to assess the impact of marijuana on the fetus is often compounded by polysubstance abuse. Investigators often must limit their work to countries like Jamaica, where women who smoke cannabis are not particularly likely to use other drugs. Alternatively, researchers can estimate the effects of other drugs like alcohol and cocaine and then see if marijuana use contributes to additional problems.

The limited available research suggests that marijuana may have little effect on offspring when they are young. Problems related to prenatal marijuana exposure may not appear until children reach the age of 4 or older. These might include an increase in problems related to attention and delinquency. Yet the women who choose to use cannabis during pregnancy may have behavioral problems themselves, making it unclear if their children develop troubles because of marijuana exposure, genetics, or poor parenting. Animal studies can avoid this problem by administering the drug to a random sample while ensuring that another group receives no drugs. These experiments show that extremely large doses of THC can lower birth weights, as well as increase spontaneous abortions and deformities, but generalizing these data to humans requires considerable caution (Zimmer & Morgan, 1997).

Research on extremely young children exposed to marijuana prenatally shows few effects. A study of more than 12,000 newborns found no link between cannabis use and gestation, birth weight, or malformations (Linn et al., 1983). Other studies found statistically significant results in large samples, but these often have little practical meaning. For example, research on 583 women showed shorter gestation periods for those who smoked cannabis 6 times per week while pregnant. Yet their babies were born an average of only 6 days early, once investigators controlled for the effects of alcohol and nicotine. Other work has found no link between marijuana use and the length of gestation (Witter & Niebyl,

1990). Cannabis use had no impact on birth weight, either (Fried, Watkinson, & Willan, 1984).

Other studies of very young children prenatally exposed to marijuana also show few meaningful effects. At 3 days and at 1 month of age, the offspring of mothers who smoked cannabis in Jamaica seemed no different from those born of mothers who never touched the drug. In fact, children of heavy users appeared less irritable, as well as more alert and stable (Dreher, Nugent, & Hudgins, 1994). In addition, the cognitive abilities of Jamaican children age 4 or 5 appeared unharmed by prenatal marijuana exposure (Hayes, Lampart, Dreher, & Morgan, 1991). These studies may provide some of the best information on the impact of marijuana exclusively because the use of other drugs is less common in Jamaica. Several North American studies also showed no effect of prenatal marijuana use on a few different measures. For example, American children exposed to marijuana prenatally showed no deficits on gross motor skills at age 3 (Chandler, Richardson, Gallagher, & Day, 1996) and no differences in total growth at age 6 (Day, Richardson, Geva, & Robles, 1994).

Although the studies above suggest little negative consequence for smoking marijuana during pregnancy, research that follows the children for a longer period reveals some potentially disturbing links to cognitive abilities and behavior problems. For example, children exposed to marijuana prenatally showed problems with a sustained attention task when they reached age 6 (Fried, Watkinson, & Gray, 1992). Another commendable longitudinal study followed over 600 mothers through pregnancy until their children reached age 10 (Goldschmidt, Day, & Richardson, 2000). Prenatal exposure to marijuana predicted several behavioral problems in this sample. Mothers who smoked cannabis while pregnant reported that their children were more impulsive and hyperactive and had more trouble paying attention. In addition, these children's teachers rated them as more delinquent. Mothers who did not smoke cannabis during pregnancy had half the rate of delinquency in their children as the mothers who smoked one joint per day. These effects remained even when the researchers statistically controlled for other contributors to these problems, including the mother's use of other drugs, her depression, and her hostility.

These data suggest that prenatal exposure to cannabis can increase troubles many years later. Nevertheless, the authors caution that they did not take the mother's own behavioral problems into account. Perhaps inattentive, hyperactive, impulsive, delinquent women are more likely to use cannabis during pregnancy. These problems might also have a genetic component. Thus, the children might have inherited these troubles from their mothers regardless of marijuana exposure. Alternatively, mothers with these qualities may serve as poor parents, leading their children to develop problems.

In addition, the researchers performed dozens of analyses only to reveal a few significant effects. They did not correct their statistics for the number of analyses conducted. Therefore, some of these findings may have occurred simply by chance. As a result, a clear, confirmed link between prenatal exposure to cannabis and later problems remains elusive. Nevertheless, pregnant women would probably do well to abstain from all drugs, as their long-term impact on offspring is often negative or unknown.

Immune Function

The impact of smoked marijuana on immune function remains a controversial topic, particularly given recent medicinal use of the drug. Many people employ cannabis in the battle against AIDS-related wasting and the anorexia associated with cancer chemotherapy. Physicians and patients show an understandable concern about any treatment that might impair the ability to ward off illness. People with AIDS and cancer need all their immune cells to function as well as possible. Research on THC and immunity has focused on individual cells, live animals, and humans. Generally, the impact of the drug is most dramatic on small cultures of cells and has less effect on immunity in people.

The vast majority of research on marijuana and immunity focuses on the white blood corpuscles (leukocytes or lymphocytes) known as T-cells. The thymus gland helps these leukocytes develop, hence, the name "T" cell. These lymphocytes can bind to cells that have been infected with a virus. The T-cells then release molecules that will destroy the infected cell. This process not only eliminates the site of the infection, but it also prevents the spread of the virus. Thus, these T-cells play an important role in immune function.

One initial study found that T-cells taken from chronic marijuana smokers were less responsive to immune system stimulants than the

T-cells taken from nonsmokers (Nahas, Suciv-Foca, Armand, & Morishima, 1974). Yet this finding has not replicated in other work. Many studies have shown no immunity problems in cells taken from marijuana smokers (IOM, 1999). Another research approach involves extracting human cells from nonsmokers and exposing the cells to large amounts of THC or cannabis smoke. This treatment decreases the cell's responses to chemicals that normally stimulate immune function. Yet heavy doses of caffeine, aspirin, and alcohol show comparable effects. Thus, these data do not justify alarm about marijuana's potential impact on immunity.

Another intriguing line of research focuses on alveolar macrophages, the immune cells that clear infectious organisms from the lungs. Researchers used a saline wash of the lungs to remove some macrophages from people who smoked marijuana regularly. Compared to the cells from nonsmokers, the macrophages extracted from marijuana smokers were less able to kill fungi, bacteria, or tumor cells. Although the behavior of extracted cells may not generalize to the immune function of humans, these data support the idea that heavy marijuana smoking could lead to pulmonary problems by inhibiting the immune cells of the lungs (Tashkin, 1999). Nevertheless, this sort of treatment of isolated cells may reveal little about the way that they might function inside a human. Thus, the relevance of this work to real people catching colds, flu, or other illness may be limited (Zimmer & Morgan, 1997).

Research from living animals might prove more relevant than findings from extracted cells. Studies of guinea pigs and mice reveal that THC can reduce their ability to ward off viruses and bacteria. For example, guinea pigs given THC and then exposed to the herpes virus showed more symptoms than those guinea pigs who do not receive THC. Mice treated with THC had less resistance to other viruses and bacterial infections, too (Cabral, 1999). Yet some of these studies use enormous doses of THC, as high as 200 mg per kilogram of body weight. An equal amount, based on weight, for a 160-pound person, would require smoking more than 700 joints. This dosage obviously surpasses the consumption of even the most devoted cannabis users. Critics of this research emphasize that these doses are unrealistic analogues of human consumption.

Given the difficulties inherent in generalizing from animal research, studies of humans may provide the most relevant evidence for THC's impact on immune function. An ideal scientific approach would randomly assign people to smoke marijuana or a placebo for months and months, and then measure which group gets sick more often. Obvious ethical problems have prevented research of this type. Instead, studies focus on the rates of illness in people who choose to use cannabis and in those who do not. This type of work must control for differences between these groups in lifestyle habits and the use of other drugs. These studies are rare. One shows a small increase in respiratory symptoms in cannabis smokers (Polen, 1993). Comparable studies of HIV-positive people found that smoking marijuana had no impact on their immune function (Coates et al., 1990; Kaslow et al., 1989). Another study actually revealed that heterosexual men who used cannabis regularly were less likely to develop non-Hodgkin's lymphoma (Holly, Lele, Bracci, & McGrath, 1999). Dronabinol, the synthetic version of THC prescribed for oral consumption, does not come with warnings about potential loss of immune function. The Institute of Medicine has concluded that the impact of marijuana smoke itself requires further research (IOM, 1999). Given results like these, THC does not appear to impair immunity, and limited ingestion of marijuana smoke may also have little effect.

Conclusions

A great deal of research addresses the role of cannabis on health. Marijuana does not appear to have a toxic dose. The drug can exacerbate symptoms of some mental illnesses, particularly psychotic disorders like schizophrenia. Yet it does not appear to cause these mental health problems. Cannabis's impact on brain structure is minimal. Nevertheless, sensitive measures of brain function reveal subtle changes associated with years of regular use. Respiratory symptoms like bronchitis and wheezing appear more often in chronic cannabis users; they also show changes in bronchial cells comparable to those seen in early stages of lung cancer. High doses of cannabinoids can alter sperm production and reproductive hormones, but these effects are temporary. The impact of THC and cannabis smoke on immune function may require further investigation, but data have yet to show that smoking marijuana increases the rates of infectious disease.

These results confirm that marijuana is neither completely harmless nor tragically toxic. Compared to other drugs that are currently legal, its impact on health is minimal. People with psychotic disorders should probably avoid cannabis. Chronic daily use obviously creates potential problems for the quick performance of complex tasks. Smoking every day undoubtedly taxes the lungs. Men attempting to impregnate women may have more luck if they abstain from cannabis. Pregnant women should probably avoid all drugs. Nevertheless, occasional use by healthy adults does not appear to create dramatic mental or physical illness. Cannabis seems to have fewer negative health effects than legal drugs, like alcohol, caffeine, or tobacco, and kills far fewer people.

8
Medical Marijuana

Proponents of legalizing cannabis for medicinal use suggest the drug could help many who currently suffer from illness and disease. Opponents of the idea assert that the legal drugs currently available provide appropriate relief from relevant symptoms. These different viewpoints have inspired spirited debates. An unbiased assessment of the drug's costs and benefits requires extensive research. Investigations must reveal the drug's ability to alleviate symptoms without creating unsatisfactory side effects. This chapter outlines a brief history of marijuana's medical uses and discusses important considerations in evaluating relevant research. An overview of marijuana's applications follows, examining the available data on its utility as a treatment for many physical ailments.

Smoked cannabis clearly helps some problems and may cost less than other medications. Synthetic cannabinoids can also alleviate symptoms of many disorders. Data suggest that cannabinoids can work well alone; they might also function effectively as part of a combination of therapies. For certain disorders, standard medications other than the cannabinoids remain the treatment of choice. Yet given the vast individual differences in reactions to medications, a few people may not improve with standard treatments and may respond better to medical cannabis. There is not enough research on most medical applications of cannabinoids to draw any firm conclusions about efficacy. Further work on marijuana's medical utility appears warranted.

A Brief History

Medicinal uses for cannabis date back to 2737 B.C., when the Chinese emperor and pharmacologist Shen Neng prescribed the drug for gout, malaria, beriberi, rheumatism, and memory problems. News of the medication spread throughout the world. The drug helped reduce symptoms in India, Africa, Greece, and Rome. Many authors assert that medical marijuana treatments would not have reached other countries unless they had meaningful efficacy. Dr. William O'Shaughnessy introduced the medication to Europe in the 1830s. By the early 1900s, some of the most prominent drug companies in Europe and America marketed cannabis extracts as cures for a variety of symptoms, including headache, nausea, cramps, and muscle spasms. Tinctures of cannabis may have had problems because of inconsistent potency, but they were often as good or better than other medications available for the same symptoms (Abel, 1980).

In the United States, the Marijuana Tax Act of 1937 discouraged medical (and recreational) use by requiring an expensive tax stamp and extensive paperwork. By 1942, against the recommendation of the American Medical Association, the *U.S. Pharmacopoeia* removed marijuana from its list of medications. This move eliminated research on the medical efficacy of the drug in this era, but recreational use increased. Users eventually noticed an impact on physical symptoms. Clinical lore about these medicinal effects spread. In 1970, the Comprehensive Drug Abuse Prevention and Control Act separated substances based on perceptions of their medical utility and liability for abuse. The act placed marijuana in Schedule I with heroin, mescaline, and LSD, making it unavailable for medical use. Despite this classification, the federal government allowed a few patients to receive the drug as part of a compassionate use program. Ideally, this program would have permitted data collection to help investigate therapeutic effects. New research on animals and humans eventually revealed medical potential for smoked marijuana, as well as individual cannabinoids.

By the early 1990s, the number of applications to the compassionate use program increased exponentially as people with AIDS sought relief from nausea and loss of appetite. The Department of Health and Human Services officially terminated the program in March 1992. Nevertheless, by the fall of 1996, California and Arizona had passed legislation permitting medicinal use of the drug. At least half of the remaining states have put forth comparable initiatives (Rosenthal & Kleber, 1999). These laws, however, conflict with federal legislation. Thus, possession of cannabis, even for medical purposes, remains a federal offense. Despite the risks, the rates of use for medical marijuana remain high. Research has continued, but only in very special circumstances, often using animals rather than human participants.

Research Considerations

Although research cannot resolve all the legal and ethical issues related to medicinal use of marijuana, it can address the drug's efficacy in treatment. Ideally, data on the utility of cannabis may inform these ethical and legal debates. Several key issues are important in evaluating research on medical marijuana. These concern the advantages and disadvantages of case studies and randomized clinical trials, as well as the relative costs and benefits of alternative medications. Case studies and randomized clinical trials each provide important information. Almost all medical uses of marijuana started with successful treatments of individual cases. One person found the drug helped alleviate a symptom and simply spread the news. Physicians published some of these reports, which occasionally inspired formal research projects. These case studies are superb for generating ideas for further work. Nevertheless, opinions vary on whether or not they provide enough information to encourage prescribing marijuana or cannabinoids. Fans of case studies emphasize that medical problems have unique features. Essentially, every use of every therapy is its own case study. Individual responses to drugs vary. As a result, physicians alter dosages and treatments based on ideographic reactions.

Proponents of case studies also mention that many medications gained widespread use based on only a few positive results, including aspirin, insulin, and penicillin. They emphasize that large studies require considerable time and expense, potentially preventing people from using a helpful drug. These arguments in support of case studies can be particularly compelling when previous research has already established a medication's safety. For example, a few studies in the mid-1970s showed that a daily aspirin might help prevent a second heart attack. Yet a large study of the treatment did not appear until 1988. Without a large clinical trial, physicians did not recommend a daily aspirin to prevent a second heart attack. This bias against smaller studies cost thousands of lives. Many people died during the lag between the initial evidence and the completion of a large clinical trial (Grinspoon & Bakalar, 1997).

In contrast, single cases also have many drawbacks. People tend to publish and remember the successful treatments but forget the failures. Without a placebo control, we do not know if improvements arose simply from expectation. Many symptoms ebb and flow with time. Perhaps some individual cases would have spontaneously recovered without any treatment. To minimize these potential problems, researchers perform randomized clinical trials. They randomly assign large samples of participants to receive cannabinoids or a placebo. If the treatment group improves more, the healing effects clearly do not stem from some natural ebb and flow in the symptoms or from a patient's expectations that the drug will work. These studies are expensive and time consuming, but they can provide the best data possible. Clinical trials of many drugs receive funding from drug companies. Yet given the limited potential for smoked marijuana to generate a profit for these companies, funding randomized control trials to establish its medical efficacy remains difficult.

Another issue important to the evaluation of medical marijuana concerns relative costs and benefits. Many evaluators suggest that cannabis must outperform all other available drugs in order to receive approval for treatment (IOM, 1999). Most supporters of this idea prefer established drugs based on the belief that they have lower potential for abuse. Physicians and patients must consider this cost relative to the drug's advantages. Critics of this idea accuse drug companies of interfering with marijuana research because of its low potential for increasing their profits (Herer, 1999). These critics highlight that the approval of other medications usually requires simple evidence of safety and efficacy, not superiority to other drugs. For example, the Food and Drug Administration (FDA) approved fluoxetine (Prozac) based on its ability to relieve depression better than a placebo. The FDA did not require data comparing it to other standard antidepressants. Thus, marijuana should only need evidence of efficacy and safety to receive approval for medical use.

In addition to established efficacy, the price of the drug and its side effects also contribute to its costs and benefits. Price and side effects play an important role in comparisons between oral THC, smoked marijuana, and other medications. Dronabinol (Marinol), the synthetic version of THC, costs as much as $13 for a 10 mg pill (Rosenthal & Kleber, 1999). (Typical treatments can require two of these pills per day.) The price of dronabinol can drop to approximately $8 for pills purchased in bulk. (A special program provides the drug to low-income patients at a reduced rate.) The same 10 mg of THC appears in half of a typical marijuana cigarette. This amount of cannabis costs less than $5 if purchased in bulk on the underground market. The price could fall markedly if the National Institute on Drug Abuse (NIDA) provided the marijuana or if the government lifted legal sanctions. Thus, smoked marijuana is cheaper, providing a clear advantage over oral THC and many other drugs.

Smoked marijuana also may have fewer side effects than oral THC and many other drugs. Patients can smoke a small amount, notice effects in a few minutes, and alter their dosages to keep adverse reactions to a minimum. Long-term health effects appear in chapter 7, but smoked marijuana for brief interventions or as a treatment for the terminally ill has no more negative side effects than many other popular drugs.

Controlled studies reveal that cannabinoids can decrease pressure inside the eye for glaucoma patients, alleviate pain, reduce vomiting, enhance appetite, promote weight gain, and minimize spasticity and involuntary movement. Other work suggests additional therapeutic effects for asthma, insomnia, and anxiety. Yet only a few studies have compared cannabinoids to established treatments for these problems. Case studies and animal research suggest that the drug may also help a host of other medical and psychological conditions. These include seizures, tumors, insomnia, menstrual cramps, premenstrual syndrome, Crohn's disease, tinnitus, schizophrenia, adult attention deficit disorder, uncontrollable violent episodes, post-traumatic stress disorder, and, surprisingly, drug addiction. The cases may provide enough evidence to stimulate researchers to conduct randomized clinical trials examining the impact of cannabinoids on these problems. The evidence of marijuana's effectiveness for treating each of these medical conditions appears below.

Elevated Intraocular Pressure

Glaucoma, the name depicting a group of problems characterized by raised pressure within the eye, affects over 67 million people worldwide. Approximately 300 people out of every 100,000 suffer from the disorder. More than 2 million Americans have glaucoma, 80,000 of whom cannot see. The heightened pressure within the eye eventually damages the optic disk, hindering vision dramatically. It is the leading preventable cause of visual impairments. Only cataracts, a currently unpreventable condition, cause blindness in more people. The prevalence of glaucoma increases with age and varies with ethnicity. The most common form of the disorder appears in 1% of people over age 60 and 9% of people over age 80. Individuals of African or Caribbean descent have higher rates. For example, over 3% of Jamaicans develop the disease. Eliminating this disorder could clearly minimize extensive financial costs and personal anguish (IOM, 1999; Quigley, 1996; West, 1997).

Treatments for glaucoma have focused on techniques for decreasing intraocular pressure to minimize damage to the optic nerve. Smoked marijuana undoubtedly lowers the pressure within the eye, as established over 30 years ago (Hepler & Petrus, 1971). At that time, the only drugs available for lowering intraocular pressure caused aversive side effects. Many patients on these medications reported blurred vision, headache, frequent urination, and racing heart. Moreover, the drugs were ineffective at lowering intraocular pressure for some people. Multiple surgical techniques developed as interventions, but not without associated risks. Synthetic THC in pill form also lowers intraocular pressure but suffers the usual drawbacks associated with oral administration. (The pills do not act as quickly as smoked marijuana. Patients report that monitoring their dosage is easier with smoked cannabis, too.) Researchers developed an eye drop containing THC, but it failed to decrease intraocular pressure.

A few glaucoma patients braved extensive bureaucratic burden to get legal medical cannabis. They turned to the government's compassionate use program before it closed in 1992. Three glaucoma patients currently receive cannabis cigarettes from the NIDA. Case studies document that marijuana has kept their intraocular pressures down and preserved their vision for many years (Randall & O'Leary, 1998).

Although smoked cannabis lowers pressure in the eye, it is not the perfect treatment for glaucoma. One potential drawback of marijuana concerns its short duration of action. Intraocular pressure creeps upward within 3 or 4 hours of smoking cannabis. This predicament forces users to smoke many times per day in order to avoid damage to the optic nerve. Some patients may not adhere to a strict regimen like this one, particularly over years and years of treatment. An alternative treatment that would only require a single dosage per day would have a meaningful advantage. This issue has become particularly important given recent crackdowns against smoking in public. Anti-smoking laws might force medical users to delay their dosages while working or traveling. The cognitive and subjective changes associated with marijuana intoxication also seem like an aversive side effect, but most patients develop tolerance to these reactions at the dosages needed to lower intraocular pressure.

After a decade of basic research, investigators in the West Indies developed Canasol, a cannabis derivative administered as eye drops that can lower intraocular pressure. Unlike the first THC eye drops, Canasol can decrease intraocular pressures up to 50% within 15 minutes. The drops are inexpensive, have no psychoactive impact, and appear to cause few side effects. They may work better when combined with other topical agents that reduce pressures (West, 1997). Despite the potential benefits of this relatively new treatment, people with years of experience using marijuana to control intraocular pressure remain reluctant to risk their sight by switching to a different medication. They report that changing drugs after years of positive experience seems unnecessary (Randall & O'Leary, 1998). New research on glaucoma treatment focuses on preserving the optic nerve and retina rather than lowering the pressures. Given the advent of Canasol and this new direction for research, the Institute of Medicine has suggested that studies of smoked marijuana will not be a priority for glaucoma research (IOM, 1999). Some patients may choose the drug if all alternatives fail, but current medications seem appropriate.

Pain

Patients seek medical assistance for pain more often than any other symptom (Andreoli, Carpenter, Bennet, & Plum, 1997). People experience a variety of pains that include diffuse, throbbing pressures or sharp, specific aches. Entire journals devote volumes to research on pain treatment. Some therapies are quite simple and cause few side effects. For example, a mere placebo can minimize pain in 16% of surgery patients (McQuay, Carroll, & Moore, 1995). Relatively simple behavioral interventions also decrease pain. Symptoms often vary with tension and mood. Thus, relaxation, stress reduction, and biofeedback can help significantly (Morley, 1997). Alternative treatments, like acupuncture, alleviate symptoms in some studies but not others, perhaps depending on the intensity and location of the pain (Kleinhenz et al., 1999; Van Tulder, Cherkin, Berman, Lao, & Koes, 1999).

Despite the success of other treatments, pharmacological interventions remain extremely popular remedies for pain. The simplest include aspirin, acetaminophen, ibuprofen, naproxen sodium, and other over-the-counter analgesics. Americans consume over 10,000 tons of these drugs a year. They are relatively cheap, have few side effects at appropriate dosages, and work well for mild pain. Nevertheless, they all can be toxic. An aspirin overdose can damage stomach lining, liver, and kidneys. A dozen acetaminophen tablets can kill a child.

Other pain killers that help severe symptoms include opiates like morphine and codeine. These work quite well even for extreme distress, inducing analgesia and an indifference to pain. People take them to recover from acute stressors like surgery. Chronic pain patients may have pumps installed in their spinal cords to release these drugs continuously. The primary drawbacks of the opiates concern their potential lethality and high liability for abuse and dependence. Opiate overdoses can be fatal. People develop tolerance quickly and often increase their doses with continued use. Withdrawal from these drugs includes extremely aversive flu-like symptoms and spastic muscle twitches (Maisto et al., 1995). Thus, alternative pain medications with fewer problems could prove extremely helpful.

An ideal analgesic would have little potential for abuse but still provide inexpensive, rapid, complete relief without side effects. No single drug has all of these qualities for treating the many types of pain. Thus, investigators have developed a multitude of analgesics. Cannabis may make a promising addition to this list. Physicians have used marijuana to alleviate pain since the beginning of the first century, when Pliny the Elder, the Roman naturalist, recommended it. The Asian surgeon Hua T'o used cannabis combined with alcohol as an anesthetic by 200 A.D. (Abel, 1980). In modern times, clinical lore and case studies support cannabis-induced analgesia. A case study reveals that oral THC can reduce phantom limb pain—the odd, aversive sensations that seem to come from amputated body parts. Another case shows that smoked marijuana can alleviate the pain of arthritis. A third suggests a tincture of cannabis can relieve tooth and gum distress (Grinspoon & Bakalar, 1997). This evidence generates intriguing hypotheses but cannot prove that effects stem from expectancy rather than genuine pharmacology. Given the dramatic impact of placebos on pain, examinations of expectancy remain extremely important. Different types of research have addressed the analgesic powers of smoked marijuana or the cannabinoids. In addition to these case studies, formal projects with larger samples also focus on this issue. These projects include tests of marijuana's painkilling effects on laboratory-induced discomfort, as well as pain from surgery, headache, and chronic illnesses like cancer.

Laboratory Stressors

Some studies examine the reactions of volunteers to aversive stimuli. Participants ingest THC and then receive electric shocks or place their fingers under hot lights or into freezing water. Initial research was not encouraging. In the 1970s, this work offered little support for cannabis as an analgesic. One study found THC actually increased sensitivity to pain. Smoked marijuana yielding approximately 12 mg of THC made people less tolerant to electric shocks (Hill, Schwin, Goodwin, & Powell, 1974). A 25 mg dose of oral THC failed to increase the threshold of pain from cold water (Karniol, ShiraKawa, Takahashi, Knobel, & Musty, 1975). The only supportive study at the time revealed that intravenous THC significantly increased the level of shock or pressure that participants first indicated as painful. Yet this last study found that the drug had no impact on the maximum amount of pain that participants could tolerate (Raft, Gregg, Ghia, & Harris, 1977).

One criticism of these laboratory studies concerns the reliability of their measures of pain. A person's threshold for pain produced by electric current is not particularly stable from one day to the next. The poor repeatability of this measure inspired the development of a new, more reliable test of pain threshold. The new test focuses on people's reactions to heat. Each participant places a finger in a specified position near a hot light bulb and withdraws it when the heat starts to hurt. A photocell detects the exact amount of time people leave their fingers near the bulb. Reactions to this test of pain tolerance vary less than reactions to shock or cold water. In short, this heat test is more reliable than the measures used before.

Participants show marijuana-induced analgesia on this heat test. In one study, they took up to 18 puffs of marijuana (3.5% THC) or placebo. The marijuana allowed them to leave their fingers beneath the lamp longer before experiencing pain. Generally, the more puffs of marijuana that the participants took, the longer they could hold their fingers under the light. These data support marijuana's analgesic effects, using a more reliable pain measure. Notably, this study also used stronger marijuana than the previous one that found no analgesia (Hill et al., 1974). This experiment suggests marijuana may reduce pain to laboratory stimuli. Nevertheless, the results may not generalize to situations more relevant to medical use. Thus, other work has focused on pain from surgery or illness, which may have many more practical applications.

Surgical Pain

A more practical approach to the study of marijuana's analgesic effects involves using the drug after surgery. Studies of THC-induced analgesia after surgery report either mixed or positive results. In one study, men who needed four molars pulled had them removed in four separate sessions under four different conditions (Raft et al., 1977). They received placebo, diazepam (an anti-anxiety medication), and two different doses of THC prior to tooth extraction. Results were mixed. This study is often cited as evidence that THC produced no analgesia (IOM, 1999). In fact, 3 participants rated the low dose as good or excellent and preferred it to the placebo; 6 others preferred placebo to THC. The high dose of THC was the least desirable of all the treatments. The results suggest that marijuana may relieve pain for a subset of individuals but not others and then only at an optimal dose. A study of pain from trauma or surgery revealed that levonantradol, a synthetic version of THC, reduced pain more than a placebo (Jain, Ryan, McMahon, & Smith, 1981). This evidence suggests that cannabinoids may show some promise in the treatment of acute pain, but tells little about the potential for handling more chronic conditions.

Headache

One recurring painful condition that may benefit from cannabis treatment is headache. Migraine, a form of headache that often includes severe throbbing accompanied by disturbed vision, chills, sweating, nausea, and vomiting, can be extremely debilitating. Bright lights, loud sounds, or pungent odors can initiate the pain. Symptoms often begin with visual disturbances like seeing flashes or auras. Then sufferers feel extreme tension and fatigue (Grinspoon & Bakalar, 1997; Russo, 1998). Eventually, a pulsing begins, sometimes on only one side of the head, where blood vessels outside the cranium dilate. These expanded arteries activate nerve fibers in the scalp, causing absolute agony. In the United States, roughly

23 million people suffer from these headaches. One-fourth of these individuals have at least four migraine attacks a month. Most of these people have their first severe headache before they turn 20. Productivity lost to migraine may cost up to $17.2 billion per year.

Treatments for this form of headache remain imperfect. Biofeedback, which trains people to use relaxation and imagery to change blood flow, has proven particularly helpful. With as little as eight sessions of proper therapy, people can learn to shrink the arteries or decrease the blood flow at the site of the pain, bringing meaningful relief to a headache (Elmore & Tursky, 1981). Several medications help alleviate symptoms for some sufferers, but fail to help 30% of people. These drugs also produce aversive side effects in up to 66% of patients. The disadvantages of these medications led some migraine sufferers to try marijuana. Physicians have prescribed cannabis for headache since as early as 1874. Advocates of the treatment protested when it was removed from the *U.S. Pharmacopoeia* in 1942 (Russo, 1998). Marijuana may have an advantage over other painkillers, such as the opiates, because cannabis not only combats headache pain, but it also inhibits the nausea and vomiting associated with migraine.

Investigators have not conducted clinical trials to support marijuana's efficacy as a headache treatment, but case reports abound. Users claim that smoking cannabis at the first sign of symptoms can combat the entire episode (Grinspoon & Bakalar, 1997; Rosenthal et al., 1997). Investigations on animals suggests that a specific brain region involved in migraine, the periaqueductal gray, contains many cannabinoid receptors. This basic research, coupled with the case reports, led the Institute of Medicine to suggest that further work on cannabinoids and migraine is warranted (IOM, 1999). Ideal studies could compare cannabis products to established medications to help verify the utility of the drug. If cannabinoids prove equally effective with fewer side effects and lower costs, they might make a superb addition to the available treatments for migraines. Combinations of cannabinoids and other medications might also prove particularly useful.

Cancer Pain

Even the most severe headaches are not as consistent as the chronic pain that accompanies many illnesses, particularly cancer. Current estimates suggest that 30% of Americans will get cancer and two-thirds of those will die from it. Cancer pain can stem from nerve injury, inflammation, or the many invasive procedures that accompany treatment. Opiates like morphine often offer little relief for these pains. They are also nauseating and notorious for their potential for tolerance, dependence, and abuse. Clinical lore from the 1970s suggested that marijuana might ameliorate some of the agony associated with this disease.

At least three subsequent studies of cancer pain provide some of the most encouraging evidence for the analgesic effects of cannabinoids. In one investigation, 10 cancer patients who had reported continuous pain received different amounts of THC on different days. Each person took a placebo on one day and 5, 10, 15, or 20 mg of THC on other days. Despite the small sample, which decreased the power to detect effects, the highest two doses of THC produced statistically significant decreases in pain. In addition, the lower doses showed a trend toward analgesia that would have been statistically significant with 4 more participants. These results proved encouraging, but the authors emphasize that subjective side effects appeared frequently. At the highest dose, a majority of the patients experienced dizziness, mental clouding, or drowsiness. In addition, half also reported feeling euphoric (Noyes, Brunk, Baram, & Canter, 1975). Patients may prefer these side effects to the pain of cancer. The relative costs and benefits of oral THC may have to be assessed on a case-by-case basis.

This first study inspired a second one with a larger sample that compared THC to codeine. Thirty-four cancer patients with chronic pain each took placebo, a low dose of codeine (60 mg), a high dose of codeine (120 mg), a low dose of THC (10 mg), and a high dose of THC (20 mg) on different days. The high doses of both drugs created statistically significant relief. Their analgesic effects did not differ from each other. The low doses failed to produce statistically significant analgesia. (Ten more participants would have brought the pain relief scores to statistical significance.) These results support the idea that THC's impact on pain compares favorably to another accepted analgesic. Yet the negative side effects were more dramatic with THC than codeine. Five patients reported aversive reactions, including anxiety, depression, and loss of control. Four others experienced obvious euphoria (Noyes, Brunk, Avery, & Canter, 1975).

A third report inspired by these two studies and animal research compared cancer patient's reactions to codeine, secobarbital (a tranquilizer), and a modified form of the THC molecule. This modified molecule contains nitrogen in a benzopyran derivative and has been dubbed "NIB." NIB produced significantly more analgesia than placebo or the tranquilizer. This result supports the analgesic properties of drugs related to THC. Many people think that cannabinoid drugs relieve pain by increasing relaxation or sedation. Yet because NIB worked better than secobarbital (the tranquilizer), it would seem that cannabinoids must have analgesic effects that do not depend solely on relaxation. The analgesic effects of NIB were comparable to codeine's. The authors of this study reported that NIB has not appeared in common medical practice because of significant side effects like anxiety and dizziness. Yet, in this sample, these effects were actually minimal (Staquet, Gantt, & Machlin, 1978).

These three studies generally support the utility of the cannabinoids as analgesics for chronic pain. Studies of smoked marijuana might also support painkilling effects for chronic conditions. Recent theories suggest that the cannabinoids might combine well with opiates to create maximal pain relief with minimal negative consequences (Fuentes et al., 1999). Research on these combination therapies could reveal an important role for cannabinoids in the treatment of pain.

Despite this supportive evidence, physicians remain concerned about cannabinoid treatments because of their side effects. A meaningful number of pain patients apparently found these cannabinoids aversive. Note, however, that these studies all used orally administered derivatives. Oral absorption of cannabinoids is notoriously slow and erratic. Controlling the dosage of these medications remains very difficult. In contrast, medical users of marijuana report improved control of dosage with smoked cannabis. Individuals can take a few inhalations of the drug, wait a few minutes, and decide if they require more or not. The quicker absorption allows for easier assessment of relief. This arrangement reportedly allows the consumption of an appropriate amount of marijuana to alleviate pain without creating depersonalization, anxiety, or other adverse side effects (Grinspoon, 1971; Grinspoon & Bakalar, 1997).

Nausea and Vomiting

Many medical conditions and treatments leave people feeling queasy and sick. Most research on marijuana and vomiting focuses on cancer. Chemotherapy remains one of the most important developments in cancer treatment. Potent, toxic chemicals attack malignant cells, eliminating some tumors and stopping the growth of many. Unfortunately, healthy tissue suffers as well. Oncologists must determine dosages carefully to minimize damage to the kidneys, heart, and other organs. These drugs produce extreme nausea and vomiting that can last for days. Although many chemotherapy patients rate the loss of their hair as their primary concern, nausea often appears as the most severe side effect of therapy. After repeated treatments, patients sometimes develop conditioned reactions, making them feel ill before chemotherapy even begins. They may grow nauseated at the sight of their physician's office or the sound of the music played in the waiting room. These side effects cause some people to miss treatments frequently or discontinue them entirely, minimizing their chances of success. Patients who vomit frequently also have trouble getting appropriate nutrition. Thus, the development of antiemetic procedures can have a dramatic impact on cancer survival rates (Grinspoon & Bakalar, 1997).

Behavioral interventions, including distraction and relaxation, help reduce nausea in chemotherapy patients (Burish & Tope, 1992). Nevertheless, most antiemetic treatments are pharmacological. Physicians knew of cannabis's ability to combat nausea and minimize vomiting at least by the 1840s, when O'Shaughnessy published the research he had conducted in India. Ancient cultures likely knew of these antiemetic effects long before (Abel, 1980). As early as the 1970s, clinical lore suggested that marijuana might help chemotherapy patients. Persuasive and emotional case studies documented the efficacy of smoked marijuana, including one from renowned Harvard professor Stephen Jay Gould (Grinspoon & Bakalar, 1997).

An initial experiment appeared in the prestigious *New England Journal of Medicine* in 1975. Twenty cancer patients who had found standard antiemetics ineffective received THC or placebo beginning two hours before chemotherapy. The patients' reports suggested that THC caused significant relief, and side effects were relatively mild (Sallan, Zinberg, & Frei, 1975). Subsequent studies consistently confirmed THC's antiemetic effects (IOM, 1999). Smoked marijuana and the synthetic cannabinoids nabilone and levonantradol also decreased nausea and vomiting for chemotherapy patients (Steele, Gralla, & Braun, 1980; Tyson et al., 1985; Viniciguerra, Moore, & Brennan, 1988). Delta-8-THC, which may create less intoxication than delta-9-THC, prevented vomiting during chemotherapy for children with a minimum of side effects (Abrahamov, Abrahamov, & Mechoulam, 1995).

Despite all this support for THC's efficacy, the drug did not gain widespread popularity because of legal issues and improved alternative treatments. Synthetic THC's long status as a Schedule II drug created extra paperwork for physicians, which may have limited its use in chemotherapy. More important, subsequent research revealed superior results with other treatments. Although THC outperformed one popular antiemetic (prochlorperazine), another drug (haloperidol) equaled THC and produced fewer side effects. A different drug (metoclopamide) has outperformed all three of these (Gralla et al., 1984). Some newer antiemetics show even more promise than metoclopamide, too. Given the strong evidence in support of these medications, most physicians would probably not prescribe dronabinol unless other drugs proved ineffective. No one has performed clinical trials comparing these drugs to smoked marijuana. Smoked cannabis may permit easier absorption and also might cost less. This research may prove extremely helpful.

Although research on smoked cannabis's efficacy as an antiemetic has not yet been completed, it has the potential to be the most cost effective option. The prices of medical treatments are extremely important, particularly in the era of managed care. Some antiemetics for chemotherapy are $400 per treatment (Kattlove, 1995), but others cost as little as $35 (IOM, 1999). The amount of dronabinol (THC) typically used as an antiemetic for chemotherapy costs $50 to $100 per treatment, depending on the amount purchased at one time. Four to six marijuana cigarettes from the NIDA would provide as much THC as the typical dosage used in this research. This amount of cannabis would cost as little as $20 in the underground market and much less if the government lifted legal sanctions. Patients report that smoking allows them to use smaller doses because rapid absorption can help them monitor the drug's impact quickly. These smoked, smaller doses translate to lower costs and fewer side effects. Opponents of medical marijuana argue that smoking increases risks for lung and throat problems, but these concerns are minimal for acute use during chemotherapy. No one has developed lung cancer from smoking marijuana for a few months.

Besides expenses, other issues support continued research on marijuana and cannabinoids as antiemetics. Individual responses to almost all medications vary widely. A subset of cancer patients may not react well to standard nausea drugs. These people might find cannabinoids or marijuana particularly helpful as an alternative. In addition, patients develop tolerance to antiemetics. Switching from one drug to another may help minimize this tolerance. Cannabinoids appear to suppress vomiting via cannabinoid receptors, whereas other drugs work in different neurotransmitter systems. Thus, occasionally using cannabinoids instead of other medications may limit tolerance to the standard antiemetics.

Because they work in a different way, cannabinoids might also make a nice adjunct when used in combination with other drugs. Smaller doses of multiple antiemetics may control vomiting with a minimum of side effects. Researchers have not yet conducted studies of these combination therapies, but they may prove promising (IOM, 1999). Given the dramatic differences in responses to drugs, as well as the potential for effective combinations of cannabinoids and other medications, treatment for chemotherapy-induced nausea requires extensive tailoring to each individual case. Further research definitely appears warranted.

Diminished Appetite and Weight Loss

In an era where some of the most famous icons of popular culture look emaciated, loss of appetite and body mass may not seem like important medical problems. Yet decreased hunger, a symptom of many illnesses and a side effect of many treatments, can lead to an inappropriate loss of weight, lowering survival rates for those with serious diseases. This loss of appetite, or anorexia, differs markedly from the psychological disorder anorexia nervosa. Anorexia nervosa typically includes a distorted sense of obesity, dissatisfaction with the size of one's body, and flagrant refusal to maintain a healthy weight. Although the term means "nervous appetite loss," the expression is a misnomer. People with anorexia nervosa still feel hunger, but they ignore it in an effort to maintain drastically low body weights (APA, 1994).

In contrast to anorexia nervosa, people who experience the symptom of anorexia show a genuine absence of appetite and a loss of interest in food. Both anorexia nervosa and anorexia can lead to malnutrition, decreased lean body mass, and significant health problems, but their treatments are markedly different. Cannabinoids offer little help to sufferers of anorexia nervosa, who simply grow conflicted and depressed as the drug enhances their desire for foods they consider forbidden. Nevertheless, marijuana and THC can combat anorexia (the symptom) successfully (Gross et al., 1983; IOM, 1999).

Anorexia typically appears in people with cancer and AIDS. From 50 to 80% of cancer patients show the dramatic loss of lean body tissue known as cachexia or wasting. This symptom appears most often in the late stages of pancreatic, lung, and prostate cancer. Cancerous cells as well as the body's immune response to them can increase anorexia and cachexia (Bruera & Higginson, 1996). In addition, the treatments for cancer and the depression that often accompanies the disease can also decrease appetite and weight.

AIDS wasting, an involuntary loss of more than 10% of body weight coupled with diarrhea or fever, decreases survival rates dramatically. This wasting weakens immune function, increasing the chances of opportunistic infection. AIDS patients are more likely to die if their weight falls 5% below ideal. Nearly every patient whose body mass drops below two-thirds of ideal weight dies within a year. Many aspects of HIV infection contribute to weight loss. Anorexia from the disease contributes to wasting. Many AIDS medications also decrease appetite, leading to further weight loss. The impaired immune system allows microorganisms in the intestine to interfere with the absorption of nutrients. Mouth and throat ulcerations also make eating difficult (IOM, 1999; Kotler, Tierney, Wang, & Pierson, 1989). Treatments for anorexia and cachexia remain imperfect, since most focus on increasing hunger in an attempt to build body mass.

People knew that cannabis enhanced appetite as early as 300 A.D. (Mattes, Engelman, Shaw, & ElSohly, 1994). Drug lore and self-report questionnaires confirmed "the munchies," an improved enjoyment of food associated with marijuana intoxication (Tart, 1971). Experimental evidence established that this increased desire for food did not arise from a placebo effect. A 13-day study of 6 healthy men living in a residential laboratory revealed that they consumed an extra 1,000 calories after smoking marijuana—40% more than after smoking placebo (Foltin et al., 1988). These results support the pharmacological explanation for cannabis's increase in appetite. Animal research has confirmed that the cannabinoid receptor CB1 plays a key role in eating. Rats injected with an endogenous cannabinoid (anandamide) ate twice as much as rats given saline. Nevertheless, anandamide did not increase eating when researchers blocked the CB1 receptor (Williams & Kirkham, 1999). Thus, cannabinoid receptors clearly play an important role in the desire for food.

Appetite increase, like every physical and psychological reaction to marijuana, shows a great deal of variation. Cannabis's stimulation of the desire for food can depend on interactions among dosage, the mode of administration, and the length of exposure. First, the dose must be appropriate. Too little of the drug will not increase appetite. Too much will actually inhibit eating. The method for administering the drug also alters its impact. Smoking may have an advantage over alternative techniques. A single dose of smoked cannabis appears to increase appetite more than a single dose of dronabinol or one marijuana suppository. Generally, smoking and suppositories have advantages over dronabinol because they can bypass the liver in the metabolism of THC. In addition, many anecdotal reports suggest that users prefer smoking. Smokers can feel the impact of the drug quickly and moderate their dose to minimize undesirable reactions. In addition to variation from dosage and mode of administration, the number of doses contributes to appetite. Multiple administrations increase appetite more in the long run than single administrations do (Mattes, Engelman et al., 1994).

Because the studies described above focused on enhancing appetites in healthy individuals, they may tell little about the impact of cannabinoids on ill people. Despite potential advantages of smoked marijuana and suppositories, the vast majority of research on medical populations involves oral, synthetic THC. Synthetic THC has helped people with Alzheimer's, cancer, and AIDS. Alzheimer's patients often refuse food. A six-week program of oral THC helped increase their weight and minimize disturbed behavior (Volicer, Stelly, Morris, McLaughlin, & Volicer, 1997). Dronabinol has improved appetite and increased weight in one study of cancer patients, too (Gorter, 1991). Most studies of chemotherapy and nausea reveal that THC can enhance appetites in cancer patients. People with HIV and AIDS have also benefited from oral cannabinoids, showing weight gain (or at least slowed weight loss), as well as improved appetite and mood.

Seventy-two AIDS patients who took 2.5 mg dronabinol twice a day showed significantly greater appetite increases than 67 others who received placebo. The difference between the groups reached significance in only 4 weeks and continued for the full 6 weeks of the study. Despite the differences in appetite, people who received THC did not gain significantly more weight after 6 weeks (Beal et al., 1995). Nevertheless, a follow-up of 94 of these patients who subsequently all received dronabinol showed that they maintained their weights for 7 months (Beal et al., 1997). Usually, late-stage AIDS patients lose weight over this much time. Other research also supports increased appetite and weight after dronabinol treatment (IOM, 1999).

These studies of oral cannabinoids suggest that they may help increase weight, yet many patients continue to use illicit marijuana instead. This practice may have risks that oral cannabinoids do not. The hazards of smoked marijuana may increase in chronically ill people. Physicians and patients have been particularly concerned about cannabis decreasing immune function, as discussed in chapter 7. Few studies have systematically examined the drug's impact on resistance to viruses, bacteria, or tumors. The impact on T and B lymphocytes, important components of the immune system, is generally small. Unfortunately, cannabis smoke does impair macrophages, the primary immune cells of the lungs. Another worry about smoked marijuana involves contamination from bacteria or fungus spores. The aspergillus fungus can grow on cannabis and cause life-threatening lung disease in anyone with impaired immune function (IOM, 1999). Yet a few minutes in a 160-degree oven or microwave may kill pathogens, perhaps minimizing this potential problem (Rosenthal et al., 1997).

Despite these concerns, anecdotal evidence for smoked cannabis remains positive. Patients claim the usual advantages of smoked marijuana, including easy absorption, minimal side effects, and lower costs. They also emphasize that their appetites increase after a single dose of smoked marijuana, whereas dronabinol may require weeks of administration before enhancing hunger. Yet no controlled studies of smoked cannabis and weight gain appear in the literature. Reports of smoked marijuana decreasing chemotherapy-induced vomiting often suggest that appetite and weight improve for cancer patients (Dansak, 1997). After many bureaucratic difficulties, Dr. Donald Abrams of the University of California in San Francisco received approval to test smoked marijuana for people with AIDS. The results of this study will help establish if inhaled cannabis smoke can increase weight without causing harm. Only large clinical trials may reveal if smoked marijuana has meaningful advantages over dronabinol and suppositories for enhancing appetite and increasing weight.

Physicians use many treatments for these symptoms besides cannabinoids. Other medications designed to combat cachexia include a variety of hormones and drugs. Megestrol acetate (Megace), a synthetic version of the hormone progesterone, has helped cancer patients and AIDS patients put on pounds. The treatment costs approximately $10 per day. It is superior to dronabinol in helping AIDS patients gain weight, and dronabinol does not enhance its effects. Thus, megestrol and dronabinol combined do not increase weight more than megestrol alone. Nevertheless, this treatment has its own drawbacks. Side effects can include difficulty breathing, impotence, changes in liver enzymes, and blood sugar problems. In addition, people may develop tolerance to its appetite-enhancing effects (Krampf, 1997; Timpone et al., 1997). Another limitation of both megestrol and THC concerns the type of weight gain they produce. These drugs increase fat more than lean body tissue. Many users of smoked cannabis report it increases their desire for sweets rather than protein-rich foods that would contribute more to their health (Tart, 1971). Thus, patients do not increase their muscle mass with these treatments. This result has inspired the search for alternative medications.

Procedures designed to increase muscular weight include the use of growth hormone, intravenous nutritional supplements, and the controversial sedative thalidomide. Recombinant human growth hormone increases muscle tissue and already has FDA approval for other conditions. Unfortunately, it costs up to $150 per day. Intravenous administration of supplemental vitamins, minerals, and electrolytes may increase lean body mass. Yet this treatment remains expensive and requires the installation of a catheter, which invariably increases discomfort and may enhance the risk of infection.

Thalidomide, the infamous sleep aid that impairs the arm and leg development of the fetus, is not currently prescribable. Yet it may inhibit the release of certain chemicals that exaggerate cachexia. Clinical trials may show that thalidomide can help increase lean body mass (Krampf, 1997). These treatments may prove superior to cannabinoids and megestrol or work well in combination with them. Despite these limitations, the cannabinoid drugs show potential promise in the treatment of anorexia and wasting. They also may help reveal a great deal about the way the cannabinoid system affects hunger and satiety.

Spasticity

Spasticity often stems from nerve problems associated with trauma and disease. Brain damage, spinal cord injury, stroke, cerebral palsy, and multiple sclerosis can all lead to this problem condition. It usually includes uncontrollable muscle flexing, loss of fine motor functioning, and associated pain. These symptoms can create considerable distress by disrupting activities in daily life. Simple tasks can become great challenges. People with severe spasticity risk choking as they eat and drink.

Uncontrollable movements at night can interrupt their sleep. Walking grows difficult or impossible, curtailing activities dramatically. Spasms and poor muscle control can also lead to urinary incontinence. The sum of all of these problems easily creates irritability, depression, and a tremendous sense of loss (IOM, 1999; Petro, 1997c).

People have known of marijuana's relaxing effects for thousands of years, but no direct references to cannabis treatment for muscle spasms appeared in ancient literature. By the late 1830s, William O'Shaughnessy had learned of cannabis's antispasmodic effects while living in India. He used the drug to treat spasms associated with tetanus and rabies (O'Shaugnessy, 1842). Less than a hundred years later, large drug companies, including Eli Lilly and Parke-Davis, marketed cannabis tinctures as antispasmodics (Aldrich, 1997). The Marijuana Tax Act of 1937 essentially forced physicians to turn to other treatments. By the 1970s, as cannabis's popularity grew, reports of its helpful effect on spasms returned. These reports inspired formal research, which focused primarily on people with spinal cord injury and multiple sclerosis.

Damage to the spine can harm the nerve pathways that extend to muscles below the site of the injury. Erratic nerve signals may lead to spastic muscle contraction and all its associated troubles. Over 15 million people in the world have spinal cord injury; over 10,000 new cases occur each year in the United States. Many arise from accidents in motor vehicles, on the job, while playing sports, or during acts of violence. The majority occur in people under age 35 (IOM, 1999).

In the first study of therapeutic cannabis in this population, researchers in a spinal cord clinic reported that 5 of 8 men who used marijuana felt it decreased their spasticity (Dunn & Davis, 1974). An anonymous survey revealed that 21 of 24 people with spinal cord injuries and spasticity reported that cannabis reduced the symptom (Malec, Harvey, & Cayner, 1982). A single-case study revealed that oral THC was superior to codeine in reducing leg spasms in a paraplegic patient with spinal cord injury (Hanigan, Destree, & Truong, 1986). A study of 5 other paraplegic patients with spinal cord damage showed that oral THC improved reflexes and muscle activity (Truong & Hanigan, 1986). Individuals with spinal cord problems also report that smoking marijuana decreases sleep interruptions and nausea (IOM, 1999).

Spinal cord injury is not the only source of spasticity. Multiple sclerosis can also lead to the symptom. This disease is related to abnormal immune function and often leads to the destruction and scarring of neurons throughout the central nervous system. Multiple sclerosis disintegrates myelin, a fatty covering that helps nerves conduct signals rapidly. This demylenation alters nerve transmission, leading to a host of symptoms, including fatigue, depression, dizziness, blindness, incontinence, and spasticity. Over 2.5 million people in the world suffer from multiple sclerosis; 90% of them develop spasticity. Many also report muscle pain and cramps. By the 1970s, clinical lore praised marijuana's potential therapeutic effects for alleviating spasms.

The initial positive reports from patients inspired more systematic research. An extensive case study revealed that one man with multiple sclerosis who could not combat his symptoms with standard medications found extensive relief after smoking marijuana. Independent physicians confirmed the improvement (Petro, 1980). The results inspired an experiment comparing oral THC to placebo in 9 patients with multiple sclerosis. These data revealed reduced spasticity as judged by a rater blind to condition. An electronic measure of muscle tension (EMG) also showed that THC had a positive effect (Petro & Ellenberger, 1981).

These initial positive results for marijuana as a treatment for spasticity associated with multiple sclerosis motivated further work. A study of 8 multiple sclerosis patients with disabling tremors confirmed that oral THC could minimize this symptom better than placebo. Five of the patients reported a subjective sense of improvement but no other measurable changes. Two others reported subjective improvement and showed an improved ability to write or hold still (Clifford, 1983). The results may have been less dramatic because of the advanced stage of multiple sclerosis and the severity of the impairment. People may benefit more in earlier stages of the disorder. Another study of a man with multiple sclerosis showed that smoked marijuana reduced spasticity and improved uncontrollable motion (Meinck, Schonle, & Conrad, 1989). This patient also reported enhanced sexual functioning—an important advantage of marijuana rarely assessed in drug studies.

As recently as 1999, the Institute of Medicine reported that studies of multiple sclerosis lacked a good animal model, limiting research. If researchers could induce the disease in animals, they could perform many experiments deemed unethical in humans. This work could help examine the role of the cannabinoid receptors in spasticity. A new technique may provide such an animal model. Chronic relapsing experimental allergic encephalomyletitis (CREAE) involves changes in immune function that lead to tremor, spasticity, and a loss of myelin comparable to the symptoms of multiple sclerosis. Extensive study revealed that THC decreases spasticity and tremor in mice with CREAE. In addition, CB1 antagonists spoil THC's ability to limit these symptoms. This work offers considerable support for the role of the cannabinoid system in movement problems (Baker et al., 2000). These animal data and the many, small-scale human studies mentioned previously justify formal, large-scale clinical trials of cannabinoid effects on spasticity in multiple sclerosis.

Involuntary Movement

Several disorders show unintentional muscle contractions comparable to spasticity, but these diseases lead to more dramatic, debilitating motion. They can create some of the same problems associated with spasticity, including difficulty sleeping and trouble with fine motor tasks. Given the large number of cannabinoid receptors in the brain's motor areas, a few studies have examined the potential therapeutic effects of cannabinoids in movement disorders, including Tourette's, Huntington's chorea, dystonia, and Parkinson's.

Tourette's syndrome, a movement disorder characterized by uncontrollable tics and vocal outbursts, may improve in response to cannabis. Three of four published case histories revealed that smoking marijuana decreased symptoms. Investigators attributed these results to cannabis's anxiolytic effects rather than a direct impact on facial tics (Hemming & Yellowlees, 1993; Sandyk & Awerbuch, 1988). Interviews with 47 people with Tourette's revealed that 13 reported using marijuana. Eleven of them (85%) said that the drug markedly improved their symptoms (Muller-Vahl, Kolbe, & Dengler, 1997). Current pharmacological treatments for the disorder include haloperidol and pimozide, two dopaminergic medications that alleviate symptoms but cause dramatic side effects. These drugs can cause debilitating sedation, aversive muscle stiffness, and dry mouth. The cannabinoids can alter dopamine's action, too, and may have advantages over these medications, but only large clinical trials can reveal the relative efficacy of these drugs.

Researchers have applied cannabinoids in the treatment of Huntington's chorea, a heritable, degenerative disease that includes rapid muscular contractions, emotional lability, and impaired intellectual functioning. The disease characteristically involves a loss of neurons in the basal ganglia, a forebrain structure rich in cannabinoid receptors. Research on rodents reveals that cannabinoids can improve functioning in these areas of the brain and decrease involuntary movement (Sanudo-Pena & Walker, 1997). Unfortunately, human research has been less promising. A single-case study showed positive effects of cannabidiol, motivating an experiment on 15 Huntington's patients given this cannabinoid or placebo. The treatment did not improve symptoms, but cannabidiol does not activate the CB1 receptor (Consroe et al., 1991; Sandyk & Awerbuch, 1988). Further work with THC or smoked marijuana would provide a better test of the efficacy of cannabinoids in treating Huntington's disease.

Dystonias, a heterogeneous group of neurological disorders that typically include involuntary muscle contractions, also may benefit from cannabis. A CB1 receptor agonist decreases dystonia in hamsters (Richter & Loscher, 1994). Cannabidiol, which does not activate this receptor, improved symptoms in 5 dystonic patients (Consroe, Sandyk, & Snider, 1986). The positive effects of both of these drugs suggests a combination therapy that supplies multiple cannabinoids may prove superior to the simple administration of oral THC.

Parkinson's, a disease known for tremor with slow movements and muscular rigidity, involves the dopamine system. An animal model of the disorder suggests that cannabinoids might help (IOM, 1999). Nevertheless, the only published study on the topic found that 5 patients who smoked marijuana showed little improvement. They did respond positively to other medications (Frankel, Hughes, Lees, & Stern, 1990). Evidence for efficacy of THC or smoked marijuana for all these movement disorders would require considerably more work. Clinical trials comparing standard medications, alternative cannabinoid treatments, and combinations of the two may show promise for adding THC, smoked marijuana, or other cannabinoids to the arsenal against movement disorders. Overall, cannabinoid treatments for involuntary movement show more promise for Tourette's and dystonia than Huntington's chorea and Parkinson's.

Seizures

Seizure, a disturbing change of consciousness accompanied by convulsions or other involuntary movements, typically stems from synchronized firing of sets of brain cells. Approximately 30 million people suffer from epilepsy, one of the most common seizure disorders. Current medical control for seizures remains ineffective for 20 to 30% of people (Petro, 1997b). The potential promise of cannabinoids in controlling seizures remains unclear. Many case studies suggest that marijuana controls seizures (British Medical Association, 1997; Grinspoon & Bakalar, 1997). Seizures caused by epilepsy may decrease in response to cannabinoids, particularly cannabidiol (Petro, 1997b). A study of 8 epileptics found improvement in seizure symptoms when they took 200 to 300 mg of cannabidiol (Cunha et al., 1980). In another study, 15 epileptics who did not respond well to standard treatment received cannabidiol or placebo in addition to their usual medications. The cannabidiol group showed greater improvement, but the response was quite variable (Carlini & Cunha, 1981). Another study of 12 epileptics using a comparable design failed to replicate this finding, but the small sample and high variability of responses limited statistical power (Ames, 1986). Only double-blind, placebo-controlled studies of larger samples can reveal the potential therapeutic effects of cannabidiol for seizure. In light of the high rates of failure for other drugs, these initial reports suggest such studies seem warranted.

Miscellaneous Symptoms

Some therapeutic applications of marijuana have generated less attention but still hold promise for alleviating symptoms and revealing important information about the function of the cannabinoid system. A few small-scale human experiments, animal investigations, and case studies suggest that cannabinoids may help treat insomnia and anxiety, decrease asthma symptoms, shrink tumors, kill microbes, and alleviate arthritis pain. Yet the latest Institute of Medicine report did not focus on this work. The data on insomnia were quite intriguing. A small sample of insomniacs fell asleep more quickly after taking THC, but side effects at the highest dose (30 mg) were excessive (Cousens & DiMascio, 1973). In another study, 15 insomniacs found that cannabidiol improved the quality and duration of their sleep (Carlini & Cunha, 1981).

In addition to its efficacy as a sleep aid, cannabidiol may work as an anxiolytic. One experiment on this topic used public speaking to induce anxiety. Ten students took placebo, cannabidiol, or an established anti-anxiety medication (ipsapirone or diazepam), before giving a videotaped speech. The speech clearly increased each participant's ratings of anxiety. Cannabidiol lowered this anxiety significantly. Its effect compared favorably to the established anxiolytics, and it created no side effects. This evidence suggests that phobics and anxious people might also benefit from cannabidiol (Zuardi, Cosme, Graeff, & Guimaraes, 1993). Nevertheless, behavioral treatments for these disorders have phenomenal success rates without requiring any medication whatsoever (Hope & Heimberg, 1993).

Asthma attacks may decrease in response to cannabinoids, but smoked marijuana obviously seems inappropriate for any respiratory problems. Asthmatics who inhaled THC through an aerosol device showed improved breathing (Hartley, Nogrady, & Seaton, 1978). Thus, cannabinoids may make a nice addition to current asthma treatments. Animal research beginning in the 1970s revealed that cannabinoids could shrink tumors (Harris, Munson, & Carchman, 1976). Recent work illuminates the complicated biochemical mechanisms behind this process (Molnar et al., 2000). A study of rats suggests that THC's analgesic effects can apply in arthritis, alleviating pain and stiffness (Smith, Fujimori, Lowe, & Welch, 1998). Other animal studies show that cannabinoids may fight inflammation, bacteria, microbes, and fungi (Kabilek et al., 1960; Turner & ElSohly, 1981).

A few other potential medical effects of cannabis have appeared in case studies but no other formal research. These include beneficial outcomes for treating menstrual cramps, premenstrual syndrome, Crohn's disease, tinnitus, schizophrenia, adult attention deficit disorder, uncontrollable violent episodes, post-traumatic stress disorder, depression, and bipolar disorder. At least one surprising case shows cannabis has helped end addiction to other drugs. All these problems have popular psychological and pharmacological interventions that do not include the cannabinoids. Yet some of the current treatments have negative side effects. They also are not effective for everyone. The case studies often portray individuals with negative attitudes about illicit drugs, who developed troublesome symptoms, struggled with standard treatments, and eventually turned to medical marijuana or oral cannabinoids for relief. These case studies certainly have many limitations, but they clearly suggest that the therapeutic uses of the cannabinoids deserve more research (Grinspoon & Bakalar, 1997).

Legal Issues

A few states have passed legislation approving marijuana prescriptions for patients. Nevertheless, possession of cannabis still violates federal laws that carry penalties of fine, imprisonment, and forfeiture of property. These federal penalties apply even in states that have passed medical marijuana laws. Thus, local authorities may not prosecute medical marijuana users, but federal authorities often do. Legal advisors recommend that any attempts to obtain marijuana for medical purposes should follow legal channels first. A number of steps may help establish medicinal need, which may augment defense if a medical user is arrested. Treatments should begin with THC in pill form, obtained through a physician's prescription. Reactions to THC in this form may prove appropriately positive, eliminating the need to use marijuana. Any reactions to the drug should appear in medical records. If synthetic THC does not alleviate symptoms, patients should apply to the Investigational New Drug program through their physicians. Although the program remains closed, evidence of these attempts may help a later defense of medical necessity. Patients should implore appropriate state agencies and ask local politicians to appeal to programs on their behalf.

If none of these steps leads to permission, patients should carefully weigh the pros and cons of medical marijuana before violating federal laws. Patients who choose to use marijuana therapeutically should report it to their physician and monitor any changes in symptoms or decreases in other medications. Under no circumstances should the medicine be shared, sold, or given away. Patients should never grow or obtain more than an adequate supply for personal use (Zeese, 1997). Given that many states increase penalties for possession of more than an ounce of cannabis, patients might consider always owning less than this amount (Margolin, 1998). This decision to use medical marijuana can prove extremely risky. The complexities of this process reveal the odd and conflicted attitudes many Americans have about the medical use of this drug.

Conclusions

The following list summarizes the efficacy of cannabinoid drugs for medical conditions:

Little evidence for efficacy
Huntington's
Parkinson's

Potential evidence for efficacy
anxiety
arthritis
dystonia
insomnia
microbes
seizures
Tourette's
tumors

Effective
appetite loss
glaucoma (alternative treatments may work better)
nauseau and vomiting (alternative treatments may work better, but they cost more)
pain
spasticity
weight loss

Therapeutic use of marijuana and cannabinoids has a history spanning over 4,500 years. Research issues related to establishing the efficacy of medicinal cannabinoids remain complex. The costs and benefits of smoked marijuana or cannabinoids can vary widely, given the range of individual reactions to drugs. Medications that may work for the vast majority of patients can have little impact on others. These idiosyncratic reactions suggest that patients and physicians can only judge the utility of cannabinoids on an individual basis.

In general, cannabinoids show promise as medicine but require a great deal of additional study. Many patients report that smoked marijuana has advantages over oral THC. Smoking permits a quick assessment of reactions and an easy modification of dosage to minimize side effects. Yet given marijuana's current status as a Schedule I drug, researchers cannot investigate its medicinal properties. Most formal medical studies investigate dronabinol, the synthetic version of THC administered as a pill.

A few consistent findings appear in the literature on medical cannabinoids. THC clearly lowers intraocular pressure associated with glaucoma, but alternative medications function equally well. Smoked cannabis and THC can alleviate pain as effectively as established analgesics like codeine. Both smoked cannabis and oral THC can also lower nausea and vomiting. Other antiemetics may produce superior effects, but they often cost appreciably more. Both smoked marijuana and THC can enhance appetite in patients enduring chemotherapy or AIDS. (Smoked marijuana has some advantages over oral THC for increasing appetite). These drugs help weight gain, too, though new, experimental medications may lead to greater increases in lean body mass.

Many case studies and a few controlled experiments suggest cannabinoids can decrease spasticity associated with spinal cord injury and multiple sclerosis. Evidence is less compelling for the treatment of other movement disorders, including Huntington's chorea and Parkinson's disease. Seizures may decrease in response to smoked marijuana or oral cannabidiol. Case studies support the medical use of cannabis for many other problems. Combination therapies that employ cannabinoids plus standard medications could have considerable potential, but researchers have not completed the appropriate studies.

Continued work on the medicinal uses of marijuana and the cannabinoids has the potential to enlighten us on the workings of the cannabinoid system. This research could also lead to improved treatments for many who suffer from numerous medical conditions.

9 Social Problems

The most heated debates about marijuana prohibition often concern the drug's role in social problems. Ideally, drug laws should minimize the negative impact of illicit substances. Arguments for marijuana's illegal status often rely on perceptions of the social problems it might cause, including decreased productivity, dangerous driving, and uncontrollable aggression. If cannabis created such adverse effects, strong penalties for its possession, sale, and use would seem warranted. Popular publications imply that marijuana's role in amotivation, reckless driving, and aggression is a proven fact (e.g., Drug Watch Oregon, 1996; Indiana Prevention Resource Center, 1998; National Institute of Drug Abuse [NIDA], 1998). Yet data reveal that cannabis plays little role in any of these social problems. Details of the relevant studies appear below.

Amotivational Syndrome

Overview

Proponents of marijuana prohibition express concern about the drug's long-term impact on motivation. Despite data to the contrary, stereotypes suggest that regular users of cannabis, particularly adolescents, transform into apathetic slugs uninterested in school, work, or any productive activity (Nahas, 1990). Researchers first identified a subset of lethargic, unmotivated cannabis smokers over 100 years ago (IHDC, 1894). Yet these data did not prove that marijuana actually altered motivation. By the late 1960s, investigators coined the expression "amotivational syndrome" to describe indifferent, listless adolescents who smoked marijuana. Yet the tacit assumption that cannabis drained their motivation was never tested.

Educators and parents grew particularly worried that marijuana destroyed ambition in the young. Case studies suggested that amotivational syndrome included poor hygiene and depressed mood, as well as a loss of energy, productivity, and drive. Authors repeatedly emphasized that the syndrome included an absence of clear goals or focused effort. Researchers suggested that repeated exposure to the drug created this condition, perhaps through a negative effect on the central nervous system (McGlothlin & West, 1968; Smith, 1968). Despite evidence to the contrary, concern about marijuana's influence on motivation continues today (NIDA, 1998). The primary problems with research on amotivational syndrome concern defining its symptoms and proving marijuana actually causes them.

Defining Amotivation

Vague definitions and varied measurements of amotivational syndrome have led to compelling critiques of the idea. Some investigators have examined employment history and educational achievement; others look at performance on laboratory tasks. Yet all claim to measure motivation or amotivational syndrome. Nearly all measurement strategies reflect stereotypically Western values about productivity. Many researchers tacitly assume that motivated people perform well in school, work hard for their employers, and persevere on laboratory tasks. Yet some of the world's most famous achievers failed in these domains. People do not share the same goals or value the pursuit of objectives in the same way. Some cultures emphasize future plans over a focus on the present. Others clearly do not. In fact, the intense pursuit of future goals may minimize enjoyment of the present moment, leading to considerable distress (Burke, 1999).

The notion of amotivational syndrome can inadvertently pathologize behaviors that many people in other cultures find fulfilling (Morningstar, 1985). One culture's amotivational syndrome may be another culture's ideal lifestyle. For example, vacation time varies dramatically from country to country, reflecting different attitudes about leisure and productivity (Robinson, 1994). In addition, motivation and achievement do not necessarily lead to happiness or increased satisfaction in life. The idea of amotivational syndrome may present a false promise that accomplishments lead invariably to happiness.

Even within Western culture, the definitions of amotivational syndrome vary considerably. There is no formal diagnosis or established list of symptoms. Most researchers employ their own unique measures of motivation, making comparisons between studies difficult. Reports usually describe amotivation as a subtle shift in priorities. Achievement becomes less important; leisure becomes more important. Sufferers purportedly have few long-term goals or no concrete plans for attaining them. They may lose the ability to concentrate, endure frustration, and participate in life. If a marijuana-induced amotivational syndrome does exist, its symptoms do not sound similar to the obvious problems associated with the abuse of other drugs. Chronic cannabis users rarely report the drastic financial, social, and occupational difficulties typical of addiction to alcohol, opiates, or cocaine. Nevertheless, if marijuana created an absence of drive, it would clearly interfere with the steady achievement stereotypically associated with the American dream.

The purported symptoms of amotivational syndrome are hardly unique to cannabis use. Clinical depression often includes the fatigue, poor concentration, and apathy typical of amotivation. This overlap suggests that a subset of depressed people who use marijuana may account for clinical observations of amotivational syndrome. Sad, unmotivated people may happen to smoke cannabis, giving the impression that the drug has created the symptoms. In fact, the links among depression, amotivation, and marijuana consumption are not particularly straightforward.

Recent data reveal that cannabis consumption has no significant association with depression in adults. A subset of people who use marijuana to cope with problems show more depressive symptoms, but it is not clear that the cannabis caused their depression. People who first tried marijuana before age 16 showed more depression later in life. Yet this relationship disappeared when the use of other drugs was taken into account (Green & Ritter, 2000). A separate study revealed that measures of motivation correlated more with depression than with marijuana consumption, even among heavy users (Musty & Kaback, 1995). Thus, depression, rather than cannabis, may cause amotivational symptoms.

Cannabis as a Cause

The idea that marijuana diminishes motivation requires the same firm evidence of association, temporal antecedence, and isolation discussed in chapter 3 with the gateway effect. Marijuana must precede and correlate with amotivation to cause it. The symptoms also must not stem from some other contributor like personality, depression, or the use of another drug. Ensuring that amotivational syndrome arises from cannabis requires experiments. Researchers can randomly assign people to receive cannabis or placebo. This arrangement ensures that everyone is equally likely to end up in the group that smokes marijuana, assuring that any identified deficits arise from cannabis rather than personality, depression, or other drug use.

In an alternative approach, participants work after smoking a placebo and at other times after smoking cannabis. This strategy, known as a "within-subjects design," ensures that all participants work both intoxicated and sober. Investigators can then compare each person's intoxicated performance to his or her own work in the absence of the drug. Under these circumstances, any identified impairment must stem from cannabis. Thus, laboratory experiments can rule out alternative explanations for marijuana's impact on motivation. This type of research requires extensive time, effort, and funding. Cannabis use over many days should produce the lethargy and lack of ambition typical of the disorder. Only a few laboratory experiments provide enough data from repeated daily exposure to provide any meaningful conclusions.

Laboratory Performance

In one of the first studies of chronic cannabis administration, researchers employed 6 men to build chairs for 70 days. They earned $2 per chair initially, but went on strike twice and raised their fees. They had periods without cannabis, and weeks when they could purchase as much as they wanted, at $.50 per joint. For 28 days, the researchers required that they smoke at least 2 joints containing a total of 17 mg of THC. Generally, the men built fewer chairs and worked fewer hours when required to consume cannabis. They also built fewer chairs immediately after they went on strike and increased their wages. The men showed no other signs of amotivation.

This study clearly supports the idea that intoxication can decrease productivity (Miles et al., 1974). Yet it is unclear if this would qualify as evidence for amotivational syndrome. Arranging for a strike to increase wages likely required motivation, organization, and drive. Making fewer chairs might reflect lower motivation, but it more likely offers further evidence that intoxication impairs performance.

In another study of chronic administration, researchers paid 30 men to stay in the hospital for 94 days. They ingested no drugs for the first 11 days, smoked cannabis for the next 64, took a break from the drug for a week, used cannabis daily for 9 more days, and then did not smoke the last 3 days. The men smoked an average of 5.2 joints per day when the researchers permitted consumption. They were paid for daily work on two different tasks. One required adding large numbers on a calculator. The other required answering textbook questions. Participants received $.10 for each correct answer on these two tasks. Acute intoxication and chronic exposure had no impact on any measure of performance. The men showed statistically comparable total responses, total correct responses, errors, and time worked throughout the 94-day period (Cohen, 1976). These data offer no support for amotivational syndrome.

These long-term studies offer little support for cannabis-induced losses of productivity. Thus, proponents of marijuana prohibition often cite other research that demonstrates decreased motivation during intoxication. One standard way to manipulate motivation in the laboratory requires offering extra cash for good performance on tasks. In one study of marijuana's effects, researchers attempted to increase motivation and performance on simple tasks by offering financial incentives. On a reaction-time task, intoxicated people did not respond to this incentive as dramatically as the people who had not smoked cannabis. Offering extra money did not motivate people to react more quickly while high, but it did speed reaction times for people who were not intoxicated. The authors emphasize that this result offers little support for amotivational syndrome. Instead, these data mean that intoxicated people do not react to standard techniques for enhancing motivation (Pihl & Sigal, 1978).

Two other studies performed in a residential laboratory revealed that intoxicated men were less likely to perform tasks that they disliked (Foltin et al., 1989, 1990). After smoking marijuana, these men spent less time on work and chores and more time on recreational activities. Popular articles often refer to these studies as evidence for amotivational syndrome. Perhaps intoxication decreases a person's willingness to work on unappealing projects, but this effect hardly parallels the directionless apathy typical of most definitions of amotivation. If these results qualify as evidence for amotivational syndrome, then most psychoactive drugs could serve as a cause. In fact, anything that might create procrastination, including watching television, could serve as a source of amotivation.

Because laboratory studies of humans offer little evidence for amotivational syndrome, critics point out that the duration of exposure was relatively brief. A couple of months of chronic smoking may not lead to any symptoms, but those who believe in amotivational syndrome charge that the disorder does not appear until later. Few people would participate in a study for a longer period, but animal research has examined the impact of 12 months of drug exposure. In this hallmark study, researchers randomly assigned 62 adolescent male monkeys to inhale cannabis smoke or a placebo. The dosage was similar to 4 or 5 marijuana cigarettes for a human. Some received smoke daily, others only on weekends, for a full year. This arrangement permitted an examination of the impact of long-term exposure to cannabis on motivation and performance.

After the year of exposure, all the monkeys performed two tasks daily for two months. The tasks were typical for primate research. One measure, conditioned position responding, provided the monkeys with a banana-flavored pellet for pressing different buttons in response to different colors. Marijuana had little impact on correct responses. The monkeys who smoked cannabis performed as well as the placebo smokers on the rate of responding and the percentage of correct responses. A subset of monkeys who had smoked every day did not complete as many trials on this task as the placebo smokers, but only during the last month of the experiment. Those who smoked cannabis only on weekends did not differ from the controls on any measure. Thus, conditioned position responding offers little evidence for amotivational syndrome, even after a year of cannabis exposure.

Proponents of marijuana prohibition might argue that conditioned position responding did not require much motivation in the first place. Essentially, the task may have been too easy to show any amotivational effects. Fortunately, the investigators also used a more difficult measure, the progressive ratio (PR) task. In the PR task, the monkeys must press a bar an increasing number of times to receive pellets. If the monkey presses the bar 3 times to get the first pellet, he has to press it 6 times to get the second pellet, 9 times to get the third pellet, and so on. The number of required presses progresses upward after each reinforcer.

Among a subset of monkeys who practiced the tasks during their year of exposure to smoke, those who received cannabis did not perform as well as those who received placebo. This deficit even appeared for the monkeys who only smoked on weekends. In contrast, the group who did not practice showed the same performance whether they had smoked cannabis or not. That is, those who smoked cannabis but did not practice the task performed equally as well as those who smoked placebo and did not practice the task. Thus, the progressive ratio task showed marijuana-related deficits in monkeys who had practiced the task during the year, but not in the unpracticed monkeys. This result may mean that monkeys who practiced while intoxicated did not learn the task as well as the monkeys who practiced while sober. The responses returned to normal within 3 months of abstinence from cannabis, when all groups performed equally well (Slikker, Paule, Ali, Scallet, & Bailey, 1992).

This study of primates shows decreased performance on a difficult exercise after a year of marijuana use, but only in those who had practiced the task during their exposure to cannabis. The drug had no impact on the easier, conditioned position responding. The investigators did not report changes in hygiene, mood, or other symptoms of amotivational syndrome. This study and some of the human research certainly confirms that intoxication can impair performance on some tasks in some conditions. Nevertheless, this seems like rather slim evidence for full-blown amotivational syndrome. Yet many critics dismiss this laboratory evidence as irrelevant. The term often implies a failure to achieve in life, not simple deficits on laboratory tasks. To further test the role of cannabis in motivation, other investigators have examined marijuana's correlation with educational and work performance. Impairments on these life tasks appear more relevant to the idea of amotivational syndrome.

Correlations with Education and Work

Surveys of associations between drug use and job or school activities lack the experimental control found in the chronic administration studies. Investigators can only assume that marijuana use causes poor performance at work or school. Alternative explanations remain equally tenable. For example, poor adjustment in work or school might lead some people to use cannabis. A third factor may account for the association, too. Depressed people might perform poorly and choose to use cannabis. People with certain personality characteristics might choose to use marijuana and make school or work a low priority. Thus, a simple association between cannabis consumption and education or work does not prove that amotivational syndrome exists. Nevertheless, the absence of an association between marijuana and achievement might undermine arguments for cannabis-induced amotivation. It is extremely unlikely that the drug causes amotivational syndrome if use and performance do not correlate. Therefore, these studies put the theory behind amotivational syndrome at risk for refutation.

School Performance

Parents and educators express understandable concern about marijuana, amotivational syndrome, and schoolwork. Research has focused on academic achievement in college and high school students. Contrary to popular belief, over half a dozen studies reveal that marijuana smokers and nonsmokers have comparable grades in college. One typical report surveyed 1,400 undergraduates, revealing no differences between users and nonusers on grades, changes in their majors, or number of colleges attended. Chronic users (those who smoked at least 3 times a week for 3 years) took more time off from their schooling but were also more likely to plan to earn a graduate degree (Hochman & Brill, 1973).

Surprisingly, at least two other studies found higher grades in the marijuana smokers than in nonsmokers (Gergen, Gergen, & Morse, 1972; Goode, 1971). Note that, despite these findings, no one has ever proposed that cannabis could help school performance. Users and nonusers also show no differences in their orientations toward achievement, their extracurricular activities, or their participation in sports. Thus, research on college students provides no support for the idea of amotivational syndrome (Zimmer & Morgan, 1997).

Although cannabis consumption in college has no link to school performance, high school students who use marijuana have lower grades and quit school more often. Cannabis smokers in high school also spend less time on their homework and miss more days of school (Kandel & Davies, 1996). At first glance, this association between cannabis and school performance seems consistent with the idea of amotivation. Perhaps cannabis destroys motivation in young teens. Yet data do not support this restricted form of amotivational syndrome, either. Most heavy users had earned lower grades prior to their marijuana consumption, suggesting cannabis could not have caused the poorer performance (Shedler & Block, 1990). In addition, high school students who smoke cannabis heavily also tend to use alcohol and other illicit substances. Once these factors are taken into account, the link between cannabis and academic performance disappears. These results suggest that drugs other than marijuana might lower grades (Hall, Solowij, & Lennon, 1994).

Marijuana probably does not cause poor school performance. Instead, the regular consumption of cannabis in high school serves as part of a general pattern of deviance. Heavy users appear more unconventional in general. They are more critical of society, less involved in church and school, and more involved in delinquent acts. They often behaved this way before they ever discovered cannabis (Donovan, 1996). Because these young people showed these qualities before using marijuana, the drug seems an unlikely cause of amotivational syndrome in high school students. Thus, depressed, unmotivated, unconventional adolescents may choose to smoke marijuana, but the drug does not appear to create their deviance. Despite this evidence, concerns about drug use in adolescents inspired the National Organization for the Reform of Marijuana Laws to recommend that only adults consume cannabis (NORML, 1996a).

Employment

Two contradictory attitudes have developed about marijuana's impact on job performance. Many people believe the drug destroys motivation and detracts from efficiency, yet others use the drug to enhance their work. Both ideas may be true, depending on the type of job involved. People who perform repetitive, simple tasks may turn to cannabis to relieve the boredom. For example, laborers in India increased their ganja consumption 50% during the harvest season (Chopra & Chopra, 1957). In Jamaica, farm hands who smoked marijuana actually worked harder than those who did not (Comitas, 1976). Perhaps marijuana makes monotonous physical labor more bearable. In contrast, jobs that require complex or rapid decisions likely suffer during intoxication (Chait & Pierri, 1992). Thus, the acute effects of cannabis on performance may vary dramatically with different jobs.

The enduring lack of initiative that defines amotivational syndrome requires more than brief changes in work performance during intoxication. Wages, hours, and employment history may serve as better indices of motivation on the job. Research performed in countries where workers frequently smoke cannabis has shown little difference between heavy users, occasional users, and abstainers. These groups had comparable forms of employment in Costa Rica and Jamaica (Bowman & Pihl, 1973; Carter, 1980). In Costa Rica, users were unemployed more often than nonusers, probably because of imprisonment for marijuana offenses. Nevertheless, heavy users had better-paying, higher-status jobs than occasional users or abstainers. People with the most stable employment smoked 15.4 joints per day. Those who changed jobs more often smoked less than half that amount, 7.6 joints per day. The unemployed smoked even less—6.2 joints per day (Page, 1983). Perhaps people with steady employment have enough experience on the job to function properly while intoxicated and enough money to afford marijuana.

In the United States, where cannabis consumption is less prevalent, the impact of the drug on wages, hours, and job turnover still does not support the idea of amotivational syndrome. Data actually suggest some positive links between marijuana consumption and work, but only for adults. One survey of more than 8,000 young adults who held a variety of jobs showed higher wages with increased use (Kaestner, 1994a). People who had smoked more marijuana in their lifetimes earned more money. Note that this correlation does not imply that cannabis consumption actually causes better pay. Perhaps people who earn more money can afford more marijuana. Another report from the same respondents revealed a negative correlation between consumption and work hours for men. Those who smoked more worked fewer hours. Yet given that their wages were higher, they may have become more efficient at work. Hours and consumption did not correlate significantly for women (Kaestner, 1994b).

Other studies of employment histories and drug use reveal that marijuana smokers do not appear to lose their jobs more often than nonsmokers, even though employers are more likely to fire users of other illicit drugs (Normand, Salyards, & Mahoney, 1990; Parish, 1989). One study of over 10,000 military personnel found that cannabis users were discharged more often (McDaniel, 1988). This result may not actually address amotivation because possession of cannabis can serve as a reason for discharge. Some of these recruits may have performed perfectly but lost their jobs because of possession. This effect did not replicate in a survey of navy recruits, which revealed that cannabis users were discharged at the same rate as others (Blank & Fenton, 1991). The respondents in these studies were all over age 18, so these data do not address amotivation in adolescents. Nevertheless, cannabis consumption does not appear to have a dramatic negative impact on wages, hours, or job turnover in adults.

Self-Perceptions of Motivation

A few studies have used the direct and intuitive approach of asking users their perceptions of marijuana's impact on their motivation. This research did not assess the many hypothesized facets of amotivational syndrome, such as lethargy, poor hygiene, and impaired social functioning. Yet these studies do reveal that a percentage of heavy users think that the drug saps their ambition or drive. Interpreting these results requires caution. Many of these participants used illicit drugs besides marijuana. They also could have suffered from unassessed conditions that undermined their energy or motivation. Nevertheless, members of every sample believe that the drug makes them less ambitious or dynamic. In one of the first studies of this kind, researchers interviewed 99 New Yorkers by phone. These people had used marijuana an average of 27 out of the previous 30 days. Eleven of these people (11%) reported reduced energy and motivation. Yet alcohol consumption was not reported, and almost half of the sample used an illicit drug other than marijuana. The report does not reveal if these 11 people who reported less energy also used other illicit drugs, but they clearly attributed their lack of motivation to cannabis (Rainone, Deren, Kleinman, & Wish, 1987).

In another study, investigators interviewed 37 Americans who claimed to have smoked marijuana at least 5,000 times. Three of these heavy users (8%) said that cannabis had a negative impact on their work because it attenuated motivation. In contrast, 7 people (19%) said that the drug enhanced their creativity and improved their work (Gruber, Pope, & Oliva, 1997). The investigators purposely excluded people who used other drugs extensively, which may explain why they found the lowest rate of reported problems with motivation.

Other researchers interviewed 268 Australians who had smoked marijuana at least 3 times per week for the previous 10 years. Over one-fifth of them (21%) felt that cannabis made them tired, unmotivated, or listless. It is unclear if they experienced these symptoms during intoxication or afterward. The use of other drugs in this sample was quite high, which may have contributed to perceptions of decreased motivation. Almost one-third (30%) of the respondents reported problematic consumption of alcohol, a sedative that lowers motivation and energy. Almost one-fourth (24%) had used an illicit drug besides cannabis in the previous month (Reilly, Didcott, Swift, & Hall, 1998). Thus, despite heavy drinking and the use of other drugs, these heavy cannabis users still reported marijuana altered their motivation. Yet the consumption of these other drugs may have sapped their drive instead.

Note that none of the studies above had a control group that could reveal if people who do not smoke cannabis also struggle with their productivity, enthusiasm, or drive. Many people feel tired, unmotivated, and low in energy without using any drugs at all. Perhaps marijuana smokers misattribute these symptoms to the drug. They may experience the natural ebb and flow of energy that all people feel, but consider this variation a result of marijuana. For example, a study of 237 students found that roughly 5% showed amotivational symptoms whether they used cannabis or not (Duncan, 1987). These results cast doubt on the idea that marijuana attenuates motivation. Instead, a percentage of people at any given time report motivational problems regardless of their drug use. Some of these people smoke cannabis and therefore attribute their lack of motivation to the drug. Yet tenable alternative sources of these problems may get overlooked because of expectancies about marijuana.

In another study, occasional marijuana users served as the control group for a sample of heavy smokers. The 44 occasional users never smoked more than 10 times in a month. The 45 heavy smokers used cannabis daily for at least 2 years. Heavy smokers also consumed more illicit drugs than occasional users, which may account for some differences between the groups. The groups did not differ in mental health, anxiety, depression, emotional control, or happiness. Yet the heavy smokers reported that marijuana was more likely to impair their motivation. The result was statistically significant, but the investigators did not correct for the large number of variables that they examined. Thus, this finding may have appeared by chance. If it did not appear by chance, then heavy users think that marijuana impairs their motivation more than occasional users. Oddly, despite the potential deficit in motivation, heavy users reported a trend toward greater life satisfaction. Again, the investigators did not correct for the large number of comparisons, so this finding may also stem from chance. Nevertheless, these results suggest that heavy users are less motivated but more satisfied with their lives (Kouri, Pope, Yurgelun-Todd, & Gruber, 1995). Perhaps they have rejected the conventional notion that motivation and productivity are essential for fulfillment.

Because the data from all these studies are correlational rather than longitudinal, they do not reveal if the heavy smokers reported poor motivation prior to ever using cannabis. Perhaps people who do not make productivity a priority subsequently choose to use marijuana. It is also unclear if the consumption of other drugs undermined productivity. Users may attribute their decreased ambition to marijuana when other drugs may have created the effect. The fact that the one study that specifically excluded users of other drugs found one of the smallest rates of motivation problems (8%) (Gruber et al., 1997) supports this idea. Mental or physical illnesses may have contributed to these symptoms, too. Nevertheless, it is clear that a percentage of people who use cannabis at high rates feel that the drug impairs their motivation.

Summary

Laboratory studies of humans and primates offer little support for amotivational syndrome. School performance does not vary with cannabis consumption in college students. High school students who smoke marijuana do worse in school. Nevertheless, most performed poorly before they used cannabis, and many used other drugs that probably contributed to their lower grades. Employment data show no links between cannabis use and lower wages, poor work performance, or job turnover. Self-reports in heavy users show that a percentage of people think cannabis impairs their drive, but consumption of other drugs or the presence of physical and emotional problems may serve as the true cause of their lack of motivation.

No studies show the pervasive lethargy, dysphoria, and apathy that initial reports suggested should appear in all heavy users. Thus, the evidence for a cannabis-induced amotivational syndrome is weak. Yet a subset of depressed users may show the symptoms of amotivational syndrome (Musty & Kaback, 1995). These people would likely benefit from cognitive-behavioral treatments for depression, which can improve mood, motivation, and achievement.

Reckless Driving

Overview

Amotivational syndrome is not the only social problem attributed to marijuana. The drug's potential role in auto accidents has also generated considerable concern. In 1997, traffic accidents in the United States numbered 16 million and caused 43,000 deaths. Comparable numbers of crashes and fatalities have likely occurred in more recent years (Bureau of Census, 1999). These statistics raise an understandable concern about impaired driving. Many drugs can increase highway mishaps. Alcohol is perhaps the most common and notorious cause of accidents. Common antidepressants, antihistamines, and tranquilizers also reduce driving skill (Riedel et al., 1998).

Cannabis intoxication clearly alters thought and memory, leading many researchers to investigate its role in highway fatalities. Popular publications imply that marijuana contributes significantly to accidents (Mann, 1985; Swan, 1994), but data do not support these conclusions. Research on cannabis and traffic safety relies on two approaches: epidemiological studies of crashes and laboratory experiments with intoxicated drivers. In general, studies reveal that marijuana has no effect on culpability for fatal crashes if a driver's age and blood alcohol concentration are taken into account. (Younger drivers who drink alcohol account for many traffic casualties.) Cannabis also does not increase the risk of accidents that cause injury. Marijuana intoxication might increase the chances of other, more minor accidents, but no data address this question.

Cannabis may not impair driving. Laboratory experiments using driving simulators and actual performance on the road reveal that motorists intoxicated with cannabis compensate for the drug's cognitive effects. They drive more slowly, leave more space between cars, and take fewer risks. Thus, current data suggest that cannabis likely does not increase reckless driving or accidents. Nevertheless, these experiments rarely focus on dangerous situations that might require rapid responses to avoid a wreck. In addition, recent work reveals that the combination of alcohol and cannabis can meaningfully increase driving problems. Given marijuana's proven ability to impair attention and rapid responses, the National Organization for the Reform of Marijuana Laws strongly urges users to avoid driving while high (NORML, 1996a). Driving after consuming alcohol, particularly in combination with cannabis, is extremely dangerous and ill-advised. Thus, users who wish to reduce the drug's harm should never operate a motor vehicle during intoxication.

Epidemiological Studies

Nearly a dozen studies from all around the globe report the frequent presence of THC in the bloodstreams of motorists involved in accidents that caused death or injury. At first glance, these results seem to support the idea that cannabis increases crashes. Yet, depending on the study, as many as 84% of these users were intoxicated with alcohol at the time. Ethanol's detrimental effect on driving is well established and seems the most parsimonious explanation for these mishaps. Analyses that exclude the presence of alcohol revealed that marijuana's impact was not significant.

For example, data from over 1,000 drivers involved in fatal accidents in Australia revealed that cannabis was present in 11% of them. Ratings of the accident reports revealed that drivers who had consumed alcohol or the combination of alcohol and cannabis were culpable more often than drivers who were free of drugs. In contrast, ratings revealed that those who used only cannabis were responsible for accidents less often than those who used no drugs at all (Drummer, 1994).

Curiously, many studies of marijuana and traffic safety found that the odds of causing death or injury were slightly lower in cannabis users than in people who had not consumed drugs (Bates & Blakely, 1999). For example, the study of Australian motorists mentioned above showed that consumers of cannabis were 30% less likely to cause accidents as drivers who had not used any drug. A study of over 300 drivers involved in fatal crashes in California focused on motorists who tested positive for cannabis but no other drug. Unexpectedly, they were half as likely to be responsible for accidents as those who were free of substances (Williams, Peat, & Crouch, 1985).

Another investigation of over 1,800 fatal crashes in the United States found that drivers who used only cannabis were only 70% as likely to have caused an accident as the drug-free group (Terhune, Ippolito, & Crouch, 1992). None of these estimates revealed statistically lower chances of accidents in cannabis users, but the consistency of these results raise interesting questions. Although no one would recommend marijuana as an aid to safe driving, perhaps the actions of cannabis users differ from drug-free drivers when they get behind the wheel. Laboratory research provides a potential explanation for these findings.

Laboratory Experiments

Another approach to answering questions about cannabis and traffic safety involves randomly assigning motorists to ingest THC or placebo before driving. This approach has several advantages over epidemiological work. Critics might argue that epidemiological studies of THC's presence in crashes may create a confounding bias. They assert that people who choose to smoke marijuana and drive may be more disinhibited or thrill-seeking than those who do not drive during cannabis intoxication. These people also may drive more poorly in general, even while completely sober. Thus, any epidemiological evidence for elevated THC rates in drivers involved with accidents may simply reflect an underlying driving deficit correlated with the propensity to smoke cannabis before operating a motor vehicle. THC may not impair driving, but poor drivers may use THC.

Laboratory experiments can bypass this problem in two ways. Researchers can randomly assign drivers to receive cannabis or placebo. This arrangement ensures that good and bad drivers are equally likely to end up in the group that smokes marijuana before driving. Random assignment assures that any identified deficits arise from intoxication rather than a biased sample. In an alternative approach, participants drive once after smoking a placebo and again after smoking cannabis. This technique, known as a within-subjects design, ensures that all the people drive both intoxicated and sober. Then, investigators can compare each individual's performance while high to his or her own performance in the absence of the drug. Again, under these circumstances, any identified impairment must stem from intoxication. Thus, laboratory experiments rule out alternative explanations for marijuana's impact on driving.

A review of more than a dozen of these experiments reveals three consistent themes. First, after smoking marijuana, users drive more slowly. In addition, they increase the distance between their cars and the car in front of them. Third, they are less likely to attempt to pass other vehicles on the road. All these practices can decrease the chance of crashes and certainly limit the probability of injury or death if an accident does occur. These three habits may explain the slightly lower risk of accidents that appears in the epidemiological studies. These results contrast dramatically to those found for alcohol. Alcohol intoxication often increases speed and passing while decreasing following distance, and markedly raises the chance of crashes (Smiley, 1986).

Additional work performed since Smiley's (1986) review confirms these effects. One recent, comprehensive paper reported four different experiments examining the impact of THC and alcohol alone and in combination. Men and women smoked joints containing 0, 100, 200, or 300 micrograms of THC per kilogram of body weight. The active doses correspond to approximately a half, one, or one-and-a-half joints for a 150-pound person. Participants drank placebos or enough alcohol to maintain breath alcohol concentrations of approximately .04%. (This dose corresponds approximately to drinking 2 beers quickly on an empty stomach for a 150-pound man.) Participants then drove in different places on separate occasions, including a deserted stretch of road, in regular highway traffic, and on city streets. A driving instructor sat beside them, rating their performance. (A second set of controls allowed the instructor to drive if needed.) These studies have advantages over research that employs driving simulators because performance in a real car in regular traffic likely generalizes to other driving situations better.

Participants performed two different driving tasks. One task, the road-tracking test, simply involved maintaining a constant speed of 90 kilometers (roughly 55 miles) per hour and staying within a designated lane. The other task, the car-following test, involved maintaining a constant distance behind a vehicle that altered its speed and acceleration. Marijuana produced two consistent effects. The drug significantly increased lateral movement within the traffic lane. That is, participants' cars weaved from side to side within the lane more after smoking cannabis than after smoking placebo. In addition, cannabis caused drivers to increase their distance from the vehicle in front of them during the car-following test. Marijuana did not alter any other way that the drivers handled the vehicle, maneuvered through traffic, or turned the car. In contrast, alcohol not only increased lateral movement in the lane, but it also impaired vehicle handling and maneuvers. The two drugs combined produced the most impairment of all (Robbe, 1998).

Thus, although traffic accidents kill thousands each year, marijuana's role in reckless driving is markedly smaller than some popular publications imply. Epidemiological research reveals that those who test positive for cannabis and no other drug do not cause accidents any more often than people who are drug free. Laboratory research shows that cannabis intoxication increases lateral motion within the traffic lane but does not impair handling, maneuvering, or turning. Obviously, no one should operate dangerous machinery of any kind under the influence of a mind-altering drug. The National Organization for the Reform of Marijuana Laws strongly encourages users to never drive during intoxication. Nevertheless, the impact of cannabis on reckless driving appears extremely small. Although traffic fatalities remain a serious social problem, marijuana appears to play a minimal role in their cause.

Aggression

Overview

In addition to concerns about loss of motivation and reckless driving, many people fear that cannabis intoxication can lead to hostility. Reviews of the cannabis literature invariably reflect writers' prejudices. Summaries of studies on marijuana and aggression may reveal these biases more than any other area of research. Interpretations of this literature are incredibly disparate. One author's evidence for marijuana's connection to violence serves as another author's proof that the drug does not cause aggression.

Interpretations of a study of murderers illustrates this point. In this research, interviews with 268 incarcerated murderers revealed that 72 of them had smoked cannabis within a day of the homicide. Of these 72, 18 claimed that marijuana contributed to the murder in some way. Fifteen of these 18 were intoxicated with other drugs at the time, too (Spunt, Goldstein, Brownstein, & Fendrich, 1994). The researchers reported these facts clearly, but interpretations of their meaning vary dramatically. One review cites this study as an example of cannabis leading to violence (Sussman, Stacy, Dent, Simon, & Johnson, 1996). Another uses it as an illustration of the rarity of marijuana-induced hostility, emphasizing how other drugs likely account for the relationship between cannabis and aggression (Zimmer & Morgan, 1997). Thus, any interpretations of data from this field require a close reading of the original studies.

People have assumed drugs lead to violence at least since the seventeenth century. Intoxication, withdrawal, and chronic use of alcohol and stimulants clearly increase aggressive acts (Kleiman, 1992). Legislators often justify drug prohibition as an effort to decrease violence. Ironically, data suggest that strict enforcement of these laws leads to a hostile underground market and a climbing murder rate (Miron, 1999). Despite evidence for increased aggression associated with other drugs, the vast majority of work shows that cannabis does not induce hostility. This research includes the standard series of case studies, correlational reports, and laboratory experiments.

Each of these research approaches has strengths and weaknesses, but the general conclusions remain the same. Direct links between cannabis intoxication and violence do not appear in the general population. A few studies show correlations between marijuana consumption and violent acts, but these links frequently stem from personality characteristics or the use of other drugs. People who are violent or who use drugs that lead to violence often also smoke marijuana, but the marijuana does not appear to cause the violence.

Laboratory studies also find no link between cannabis intoxication and violence. Most people who ingest THC before performing a competitive task in the laboratory do not show more aggression than people who receive placebos; occasionally, they show decreased hostility. Numerous scientific panels sponsored by various governments invariably report that marijuana does not lead to violence (Zimmer & Morgan, 1997). Yet two studies reveal small but significant links between cannabis and aggression with very select populations under extremely circumscribed conditions. If these findings replicate, further work may reveal a great deal about aggression in general and subsets of individuals susceptible to provocation during marijuana intoxication.

Historical Precedent

Nearly every discussion of cannabis and aggression begins with legends about the assassins. Hasan, the leader of an unorthodox Muslim sect in the year 1090, allegedly kept his power by commanding his followers to assassinate his rivals. The fierce fighting of Hasan's devotees inspired tales of their unparalleled loyalty. Their loyalty allegedly stemmed from a belief that completing their missions guaranteed entry into paradise. One tale revealed that new initiates of the sect were drugged, blindfolded, and taken to a lush garden filled with exotic diversions. They left the garden with the promise that they would return at their deaths if they followed Hasan's orders. This experience purportedly motivated the followers to act as instructed. Later versions of the tale implied that the drug used to sedate them was hashish. Subsequent adaptations suggested that they took hashish to whip themselves into a frenzy immediately before the murders.

Oddly, these tales did not lead people to believe that cannabis aided sleep. Instead, the idea spread that hashish intoxication caused aggression. Some people asserted that the name given these murderous followers of Hasan even derived from the name of the drug. These hashish-eating killers were called "assassins." Better evidence suggests "assassin" may have originally meant "follower of Hasan." The root word "hassa" actually means "kill" or "exterminate," revealing that assassin probably means "killer" rather than "hashish eater." Nevertheless, the connection between assassins and hashish remains in the minds of many (Casto, 1970). Harry Anslinger, the first head of the Federal Bureau of Narcotics, cited the story of the assassins as evidence of marijuana-induced violence (Bonnie & Whitebread, 1974). Modern authors still suggest that the drug leads to hostility (Schwartz, 1984). This belief may stem from poor interpretations of individual cases.

Some of the most sensationalistic, gory case studies came from the Bureau of Narcotics in the 1930s. Most told of marijuana users who committed heinous crimes. Many times the details did not reveal if the crime actually occurred during marijuana intoxication. Yet media attention focused on marijuana's link to violence. Unfortunately, plausible alternative explanations did not receive the same level of enthusiastic coverage. A classic example concerned a Florida murder case from 1933. Victor Licata, a known cannabis user, killed his parents and three siblings. A local paper attributed the murders to the drug, and Harry Anslinger used the case as an example for many, many years.

Despite initial reports of this event, further investigation revealed that Licata may have heard voices at the time of the murders. He suffered from a serious, psychotic, mental illness. Many members of his family also struggled with psychotic disorders. Licata may have had a history of violence prior to his drug use. Yet none of these possibilities appeared in the press (Kaplan, 1970). A close look at another case study frequently cited by the Bureau of Narcotics revealed that the murderer had claimed to use marijuana when, in fact, he had not (Bromberg, 1939). Some authors accuse Harry Anslinger of focusing on tales like these in an effort to justify a larger budget for the Bureau of Narcotics. Others also suggest that William Randolph Hearst published anti-cannabis stories in all of his newspapers to keep hemp production from undermining the value of the forests he owned (Herer, 1999; Sloman, 1998).

Crime

A more scientific way to investigate marijuana's link to violence appeared in studies of crime rates. Researchers have looked for an association between violent crime and cannabis consumption for at least 70 years. This association does not prove that marijuana causes aggression, but any theory linking cannabis and violence would suggest that the two should covary. Early studies of military personnel, arrestees, and patients in mental hospitals revealed no relationship between cannabis and violent crime.

One typical study examined rates of aggressive crime in military prisoners. Marijuana users were no more likely to commit crimes of violence than nonusers (Bromberg & Rodgers, 1946). Some studies revealed fewer antisocial behaviors in cannabis smokers than in users of other drugs (Abel, 1977). Later research confirmed these findings. For example, a study of 109 delinquent boys revealed that violent offenses had no link with cannabis consumption, but had significant associations with cocaine and amphetamine use (Simonds & Kashani, 1980).

A few recent studies reported small but statistically significant associations between marijuana consumption and violence in select groups of adolescents. Yet the effects were extremely small, meaning that the amount of violence increased only a little as the amount of cannabis consumption increased a lot. (Correlations were approximately .20 and only reached statistical significance because of the large sample sizes). These studies asked teens about their marijuana use, as well as the frequency of their aggressive acts, but failed to assess if they were high when they were hostile. Thus, they do not support the idea that cannabis causes violence. Instead, a subset of teens may choose both to use marijuana and behave aggressively because of an underlying personality characteristic or tendency (Sussman, Simon, Dent, Steinberg, & Stacy, 1999; White & Hansell, 1998; White, Loeber, Stouthamer-Loeber, & Farrington, 1999). People who seek thrills or have trouble inhibiting themselves might engage in both cannabis consumption and violent behavior. Yet neither one caused the other. The use of other drugs, including alcohol, may be a more likely explanation for the aggression. In fact, when one group of researchers included previous violence and alcohol consumption in their analyses, the links between marijuana and aggression disappeared (White et al., 1999).

Other studies suggest that these small links between cannabis consumption and hostility do not mean that marijuana intoxication leads to aggression. For example, a group of adolescents charged with violent crimes reported that cannabis was likely to decrease aggressiveness (Tinklenberg, Murphy, Murphy, & Pfefferbaum, 1981). Fewer than 4% of people report that they think marijuana makes them angry or hostile (Davidson & Schenk, 1994; Halikas, Goodwin, & Guze, 1971). Research participants have lower scores on questionnaires designed to assess hostility, anger, and aggressiveness if they answer after smoking cannabis (Abel, 1977). Yet some of the most compelling evidence that the drug does not increase hostility stems from laboratory work that actually measures belligerent behavior.

Laboratory Research

A sophisticated way to examine marijuana's impact on aggression requires providing THC to participants in the laboratory. Few people behave in a hostile fashion in such a formal setting, so most studies provoke participants to see if they will aggress in response. A popular paradigm uses a competitive game. The participant competes against an opponent to provide a faster, correct response. The winner of each trial can give the loser an electric shock. (A later version of the task allows the winner to take money or points from the loser).

In fact, the opponent is bogus and the results are fixed. The participant loses a specified number of times. The experimenter makes it seem as if the opponent provides increasing or heavy penalties in an effort to provoke aggression. This paradigm may seem an absurd analogue of hostile interactions in everyday life. Yet former prisoners with histories of aggressive acts do behave more aggressively in this game. Frustration, drug withdrawal, and other conditions that should increase violence also increase aggression in the game (Cherek, Moeller, Schnapp, & Dougherty, 1997). Laboratory studies using this paradigm find that marijuana intoxication rarely heightens hostile responses. Participants gave stronger shocks when intoxicated with alcohol, but THC had no impact. A high dose of THC actually lowered aggression, despite the provocation inherent in the task (Myerscough & Taylor, 1985; Taylor et al., 1976). These results suggest that cannabis intoxication does not increase aggression in a normal population.

One study using a variation of this paradigm has received considerable attention because it appears to reveal increased aggression during marijuana intoxication. Eight inner-city men who regularly used cocaine and other drugs participated. Seven of them were diagnosed with antisocial personality disorder, a problem formerly known as sociopathy, which frequently accompanies troubles with drugs and violence. This study used the revised paradigm that allows participants to take away points that could be traded for cash. (Researchers dubbed this procedure "the point subtraction paradigm.") These antisocial participants showed more aggression while intoxicated with cannabis, but only for the first hour after smoking (Cherek, Roache, Egli, Davis, et al., 1993). Analyses of later sessions of the game do not appear in the article, presumably because they showed no significant effects.

Interpreting these limited effects in such a small sample proves difficult. Perhaps a subset of individuals react more aggressively after smoking cannabis. Other drugs also seem to induce greater violence in subsets of people. For example, data suggest that men with antisocial personality disorder show greater increases in aggression after alcohol than men without the disorder (Moeller, Dougherty, Lane, Steinberg, & Cherek, 1998). Nevertheless, this single, small laboratory study should not lead people to believe cannabis causes violence, particularly given the other studies that show no marijuana-induced aggression.

One other laboratory study examined aggression associated with marijuana withdrawal. The researchers used the point subtraction paradigm in a sample of 19 people who had smoked cannabis at least 5,000 times. These participants met criteria for substance dependence. The control group consisted of 20 people with markedly less involvement with marijuana. They all played the point subtraction game on days 1, 3, 7, and 28 of an inpatient stay in the detoxification unit of a hospital. Participants who were cannabis dependent behaved more aggressively than the controls on days 3 and 7. They were also more aggressive than they had been on day 1. By day 28, their aggressive responses returned to baseline and did not differ from the aggression shown by the controls (Kouri, Pope, & Lukas, 1999). This study provides an intriguing interpretation of other links between marijuana and aggression. Although intoxication does not lead to hostility, periods of withdrawal might. Perhaps the small links between cannabis and aggression in studies of crime arose because of withdrawal rather than intoxication. A replication of this study would prove most illustrative.

Summary

Despite ancient tales and widespread misperception, marijuana intoxication does not lead to aggression in the general population. Self-reports of experienced users suggest that the drug makes them feel mellow and calm rather than hostile and unfriendly. Research on crime reveals little impact of cannabis on violence. The vast majority of laboratory research shows that marijuana intoxication does not increase hostile responding. A few weak associations between cannabis and aggression arise in small subsets of the population, like delinquent teens, psychopaths, and marijuana-dependent individuals experiencing withdrawal. The drug's absence of an impact on hostility has led every major commission report to conclude that cannabis does not increase aggression.

Conclusions

Prohibitionists suggest that marijuana creates meaningful social problems, including amotivational syndrome, reckless driving, and aggression. Research in each of these domains reveals that these concerns are unfounded. Evidence for a marijuana-induced amotivational syndrome is lacking. A subset of depressed users may have inspired a few case studies that report apathy, indifference, and dysphoria, but cannabis likely does not cause these symptoms. The drug does not correlate with grades in college students. High school students who use marijuana have lower grades, but their poor school performance occurred prior to their consumption of cannabis. Cannabis users do not show worse performance on the job, more frequent unemployment, or lower wages. In addition, long-term exposure to cannabis in the laboratory fails to show any meaningful or consistent impact on productivity.

Links between cannabis and reckless driving are also weak and usually stem from co-occurring alcohol consumption. People with THC but no alcohol in their blood do not have higher rates of culpability for traffic accidents than drug-free drivers. Laboratory experiments that administer THC and placebo to motorists reveal an increased weaving within the lane that accompanies intoxication. Yet these drivers also spontaneously slow their speed, increase their following distance, and rarely attempt to pass other cars. In contrast, alcohol, even at relatively low doses, clearly impairs driving.

The association between cannabis intoxication and aggression is also unlikely. Most studies of violent crime show no link to marijuana use or small correlations that suggest a few aggressive people also happen to smoke cannabis. Laboratory research on general samples shows no increases in aggression during intoxication. Concerns about people's productivity, impaired driving, and hostility are certainly important, but altering marijuana consumption will likely have little impact on these social problems.

10
Law and Policy

The brief history of marijuana laws in the United States reflects considerable controversy. Proponents and opponents of current cannabis prohibition make moral and practical arguments for their positions. Recent efforts to decriminalize the drug have fueled debates about the implications of limiting penalties for possession. Some people view current punishments as inappropriate, given the limited negative consequences associated with marijuana use. Others long to maintain the status quo or request tougher sanctions in hope of decreasing harm and creating a drug-free America.

Several authors propose steps beyond decriminalization to legalization in an effort to eliminate the underground market in cannabis. These antiprohibitionists have suggested a variety of plans, ranging from an unregulated free market to a highly taxed, controlled, and licensed arrangement. Proponents of decriminalization and legalization suggest that changes in current laws could save taxpayers money, decrease the potential for violations of civil rights, and still keep marijuana-induced harm to a minimum. In contrast, prohibitionists assert that changes in policy would convey tacit approval of drug use, leading to increased consumption of marijuana and other illicit substances and exacerbating negative consequences.

A Brief History of Marijuana Legislation

Before the 1900s, cannabis products were legal in the United States. Although the young American Fitz Hugh Ludlow described cannabis intoxication in *The Hasheesh Eater* in 1857, few other U.S. residents had any exposure to the recreational use of the drug for many years. The practice of smoking marijuana, often attributed to immigrants from the West Indies and Mexico, had little impact on national policy before the 1930s. State and city regulations developed against the drug first. Local ordinances against "loco weed" appeared in El Paso as early as 1914. All of Texas prohibited the drug by 1919. Thirty-two states enacted marijuana prohibition by 1933, often based on stories of the drug inciting immigrants to violence.

With the help of Harry Anslinger, the first head of the Federal Bureau of Narcotics, the federal government passed the Marijuana Tax Act in 1937. This tax regulation did not make the drug illegal, but required a prohibitive fee of $100 per ounce for the transfer of marijuana. Possession of the drug without the appropriate tax stamps was a federal crime. Anslinger justified the law with graphic reports of murder and mayhem that the drug supposedly induced. Current data prove that marijuana intoxication does not lead to these crimes (see chapter 9). By 1940, every state had outlawed the drug. Public opinion held that this intoxicant could prove more dangerous than heroin. Thus, possession of cannabis carried identical penalties to heroin possession. Sanctions increased in the 1950s (Bonnie & Whitebread, 1974; Weisheit, 1992). At that time in Georgia, a second conviction for sale to a minor carried a death sentence (Himmelstein, 1986).

Attitudes changed by the late 1960s. As more young people experimented with the drug, perceptions of its effects shifted. People questioned previous media portrayals of marijuana enslaving all users and transforming them into deranged, criminal freaks. Arrests remained frequent, but penalties decreased by the mid-1970s (Brown, Flanagan, & McLeod, 1984). By 1978, at least 11 states had decriminalized possession. This decriminalization minimized the state penalties for owning cannabis, but federal laws still applied. Thus, federal authorities could still prosecute anyone found with marijuana in these states.

Many other states did not decriminalize but often dropped charges for first-time offenders possessing small amounts. A dozen states erased the record of first-time possession offenders after a period of appropriate conduct. President Carter even recommended federal decriminalization, emphasizing that the laws provided worse negative consequences than the drug. Activists in the era predicted that most states would legalize the drug within a few years (Sloman, 1998).

Despite the hopes of marijuana activists, the pendulum swayed back toward criminalization by the early 1980s. Perhaps as a result of outside political pressures, the Drug Enforcement Administration (DEA) depicted marijuana use as the United States' most serious problem (Koski & Eckberg, 1983). Penalties increased again. Some states that had previously decriminalized possession reinstated sanctions. Enforcement of paraphernalia laws grew more frequent. Possession of water pipes, roach clips, and anything else related to drug administration remains a crime. Censors attempted to classify promarijuana publications as paraphernalia, making books and magazines that discuss the drug illegal. The DEA developed marijuana eradication programs, which searched specifically for cannabis fields to burn. The programs expanded from Hawaii and California to include over 40 states. Growing concerns about the impact of drugs on the productivity of workers motivated an increase in drug testing. Support for decriminalization waned and the enforcement of legal sanctions increased (Brown et al.,1984; Weisheit, 1992).

In contemporary policy, almost anything involving marijuana carries penalties in the United States. Possession, transportation, cultivation, sales, offering to sell, and driving under the influence all qualify. Selling oregano or other legal substances as if they were cannabis is also a crime. Possession of marijuana paraphernalia also violates criminal codes. Penalties vary dramatically from state to state and increase with repeated offenses and larger amounts of the drug. Some penalties are relatively small. A first offense of possession of less than an ounce in California can lead to as little as a $100 fine. Others are markedly larger. In Rhode Island, possession of over 5 kilograms can lead to a $1,000,000 fine and life imprisonment. The potential exposure of minors to the drug also increases penalties. Many states have added sanctions for possession near a school or housing project; some states double penalties for sales to a minor.

Several states also suspend the driver's license of anyone convicted of a marijuana crime, even if the crime did not involve a car. Some areas will only reinstate these suspended licenses after treatment for substance abuse. These laws essentially mandate therapy for anyone who possesses marijuana. Suspensions may last as long as 5 years. Drug seizure laws essentially add more penalties. Police can force people suspected of marijuana offenses to forfeit cash, cars, houses, boats, farms, or any other property that may have facilitated a crime. Any property potentially purchased with money obtained through the sale of marijuana is also subject to forfeiture (Boire, 1992). A majority of people who forfeit property actually may never face criminal charges (Schneider & Flaherty, 1991). Thus, police suspicion alone may lead to the loss of a home, car, or boat.

Arguments about Prohibition

Proponents and opponents of marijuana prohibition argue about the costs and benefits of relevant laws on a number of grounds. Most arguments concentrate on estimates of expenses related to enforcing the laws and the price of drug-induced harm. Different perceptions of these factors lead to different opinions about marijuana policies. The arguments usually attempt to estimate the advantages and disadvantages of the current laws relative to alternative proposals. Ideal policy would eliminate the negative consequences of the drug cheaply and efficiently. Most arguments are internally consistent given a specific set of assumptions. Debates between proponents and opponents of prohibition often arise because they cannot agree on the same values for their underlying assumptions.

A taxonomy of prohibition arguments has developed, describing sets that ostensibly rely on morals and rights, or costs and benefits. The distinction among these types of arguments can be artificial. Assertions related to morals and rights can reflect perceptions of good and evil that purportedly transcend simple assessments of expenses or harm. Yet explanations of why some acts are wrong often rely on their associated negative consequences. Utilitarian arguments sometimes appear to weight costs and benefits in an objective manner. Yet they assign these weights based on underlying perceptions of right and wrong that often reflect a sense of morals or rights.

Purportedly Moral Arguments for Prohibition

Any discussion of morality can inflame people. These issues remain complex, emotional, and difficult to summarize. Entire books devote hundreds of pages to morality and law related to drug policy (e.g., Fish, 1998). Only highlights of some of the most prominent debates appear here. Moral arguments in support of prohibition focus on the perception of ethical behavior. Some proponents of these moral arguments claim that their rationales are independent of the consequences of actions. That is, some behaviors may be wrong, even if they do not necessarily lead to harm. Thus, these arguments do not rely directly on estimates of the damage marijuana causes. According to some legal moralists who support prohibition, cannabis remains outlawed because it is wrong. If a medication or therapy appeared that could counteract any potential harm marijuana might cause, use would remain wrong. Even in this harmless condition, the drug should continue to be illegal.

Former drug czar William Bennett uses moral explanations in his work. "The simple fact is that drug use is wrong. And the moral argument, in the end, is the most compelling argument" (Bennett, 1991). The reasons marijuana consumption is wrong often rely on incontestable ethical insight (Husak, 1998). Yet when pressed to explain these ethics, moralists often turn to utilitarian assessments of harm. For example, Barry McCaffrey, the former drug czar, argues that drugs are wrong because they are "destructive of a person's physical, emotional, and moral strength" (Raspberry, 1996). Data on the impact of marijuana on moral strength remain unavailable. Physical and emotional effects are well documented, but relying on these data may turn a moral argument into a utilitarian assessment of costs and benefits.

These moral arguments in support of prohibition may stem from attitudes about pleasure, productivity, intoxication, and self-control. One assumption underlying moral arguments concerns the idea that pleasure should only reward contribution to society. Essentially, pleasures should follow concerted, responsible productivity. Thus, consuming the drug is morally wrong because it creates pleasures that some people do not believe are properly earned. In addition, attitudes about intoxication and its link to productivity may also underlie these arguments. Any state of impaired thought may hinder productivity, which violates the work ethic many people view as intrinsically American. Others suggest that intoxication destroys the ability to behave in safe, conscionable ways. Legal moralists who support prohibition assert that any change in drug policy would be immoral because it sends the wrong message to citizens. Some authors assert that moralists think that the drug should remain illegal at any expense (Husak, 1992). Note that these arguments eventually resort to utilitarian assessments of potential harm. Explanations for the morals often lead to evaluations of costs and benefits.

Purportedly Moral Arguments against Prohibition

Some arguments against prohibition stem from perceptions of constitutional and human rights. These rationales also often rely on perceived links between rights and morality. Like the moral arguments to support prohibition, these arguments against prohibition purportedly do not rely on estimates of harm. Most focus on the guarantees of the Constitution, including the rights to freedom of religion, privacy, and property. Moralists against prohibition argue that current drug laws infringe on these rights. They assert that prohibition is immoral because it violates these principles of the Constitution.

Arguments in support of religious rights related to drug consumption appear particularly complicated. The quest for freedom of religion drove many Europeans across the Atlantic in the first place. At least two religions have formal histories of using marijuana as a sacrament: the Brahmakrishna sect of Hinduism and the Ethiopian Zion Coptic Church. Both have a long tradition of cannabis rituals. Moralists against prohibition argue that members of these churches should continue their practices as part of the religious freedom guaranteed by the First Amendment. Most draw parallels between cannabis rituals and the religious use of wine by American Jews and Christians. These supporters obviously view the benefits of religious freedom as more important than the costs of marijuana consumption.

The U.S. courts do not support these religious arguments against prohibition (*Leary v. U.S.*, 1967; *Olsen v. D.E.A.*, 1989). Prohibitionists emphasize that wine is not consumed to the point of intoxication in most religious rituals. Moralists against prohibition often point to the Jewish practice of drinking to intoxication on the holiday of Purim. They also emphasize the sacramental use of peyote in the Native American Church, which was protected by the American Indian Religious Freedom Act Amendments of 1994. Peyote clearly causes intoxication. Prohibitionists argue that peyote is used less widely than marijuana and may prove easier to keep under control. The Native American Peyote ritual also specifies particular and infrequent times for ingesting the substance. In contrast, sacramental consumption of cannabis often occurs many times per day. Moralists against prohibition emphasize the right to freedom of religion over any aspects of controlling the drug. (Note that these arguments eventually resort to utilitarian assessments. Thus, moralists against prohibition may see threats to religious freedom as more harmful than any negative consequences of peyote or cannabis consumption.)

Despite protest, the Court continues to support prohibition, and religious use of cannabis remains illegal. The hallmark Supreme Court case related to this issue, *Employment Division v. Smith* (1990), concerned sacramental peyote use. The case has generated commentaries that could fill a small library. Like all Supreme Court cases, this one does not lend itself to an easy summary. A private organization fired two substance abuse counselors for their sacramental use of peyote. When the men applied for unemployment, the state turned them down because they had lost their jobs due to misconduct. The Court ruled that the Free Exercise of Religion Clause of the First Amendment does not bar the "application of a neutral, generally applicable law to religiously motivated action." The ruling allowed the state to deny unemployment payments. This case suggests that people cannot sidestep laws for religious reasons, unless the laws unconstitutionally attempt to regulate religious practice.

Moralists against prohibition also view the ingestion of marijuana as a personal act protected by their interpretation of the Constitution's right to privacy. A huge legal literature exists on the right to privacy. Moralists against prohibition assert that, although the Constitution does not guarantee a right to privacy directly, any activity conducted alone or among intimates (that does not harm others) might have constitutional protection. This interpretation of the right rests on a few previous cases. For example, the right to use birth control has been protected under the right to privacy. Moralists against prohibition assert that cannabis consumption should qualify under this right as well. Alternative arguments suggest that the right to privacy only applies in important, fundamental decisions similar to having children. These interpretations imply that using marijuana lacks the fundamental import to fall under a right to privacy.

This right to privacy became particularly relevant in a classic state case in Alaska, *Ravin v. State*. In an effort to test the applicability of a right to privacy, attorney Irwin Ravin arranged to have himself arrested for possession in 1972. State judges determined that the right to privacy applied in this case. They emphasized that the noncommercial, individual aspects of the situation made the ingestion of marijuana consistent with the right to privacy. After this decision in 1975, the Alaska legislature removed criminal penalties for possession of up to 4 ounces of marijuana for personal use. Nevertheless, in 1990, the drug was again criminalized in Alaska. Enforcement of the new law may be rare (Gordon, 1994). A subsequent federal case (*NORML v. Bell,* 1980) was unsuccessful in arguing that the right to privacy applied to marijuana possession. The court asserted that using marijuana was not a fundamental or established right important enough to qualify for privacy protection. Thus, moralists against prohibition who rely on arguments related to the right of privacy currently have no support from the federal courts.

Another moral argument against prohibition focuses on the right to property. Thomas Szasz, the psychiatrist who gained notoriety for his debates about conceptualizations of mental illness, emphasized that drugs are personal property. Given this fact, they therefore qualify for constitutional protection. The argument suggests that any state interference with drugs violates this right to property as depicted in the Fourteenth Amendment. Szasz and other libertarians do not advocate drug consumption. They view the decision to use drugs as part of an individual's liberty and responsibility—a moral issue outside the realm of legislation.

The U.S. Supreme Court did not uphold this type of argument (*Crane v. Campbell,* 1917). In this case, an individual asserted that the state's prohibition against possession of alcohol conflicted with the Fourteenth Amendment's declaration that no State shall "deprive any person of life, liberty, or property without due process of law." The Court did not see the right to possess liquor as a fundamental privilege that no State could violate. Thus, the same rules apply to possession of marijuana.

Another argument suggests that drug use relates to the right to self-determination. Self-determination may fall under the Ninth Amendment, which emphasizes that citizens have rights that are not specifically listed in the Constitution. According to this line of reasoning, the right to determine what enters one's own body qualifies as this sort of self-determination. Legal scholars opposed to prohibition argue that this amendment applies to the possession and ingestion of drugs. Drugs were freely available to everyone at the time the framers drafted the Bill of Rights. Szasz and others suggest that the originators of the Constitution likely viewed the freedom to ingest whatever one chooses as too intuitively obvious to mention as a specific right (Szasz, 1992). Other students of history do not agree with this suggestion. Marijuana cases related to the right to self-determination have not appeared. Most arguments that rely on interpretations of the Ninth Amendment do not fare well in the courts.

Arguments Based on the Consequences of Drug Use and Drug Laws

Many prohibitionists assert that legislators developed marijuana laws to minimize any potential harm that users may cause for themselves or others. Yet data suggest that drug prohibition and strict law enforcement creates an underground market rife with violence. For example, prohibition and strict law enforcement leads to increases in murder rates (Miron, 1999). Other authors argue that marijuana prohibition evolved from racist attitudes against immigrants who used the drug (Musto, 1999). Financial incentives also may have motivated the legislation. For example, Herer (1999) suggests that William Randolph Hearst wanted to eliminate hemp production so his extensive holdings of wooded land could serve as the sole source of pulp for the production of paper. Hearst's many newspapers published alarmist tales of atrocious, marijuana-induced crimes to inspire antimarijuana legislation. The laws also made hemp illegal, allowing the tycoon's logging industry to flourish.

Whatever the origin of the laws, prohibitionists frequently point to low levels of use and problems as signs of success of marijuana control. About one-third of American adults have tried marijuana in their lifetimes, but only 3% report using the drug once a week or more (SAMHSA, 1997). Generally, fewer than 10% of regular users experience problems related to the drug (Weller & Halikas, 1980; see chapter 2). Cannabis causes markedly less harm than other drugs, particularly alcohol and nicotine. Few people clamor for admission to drug treatment for marijuana troubles. No one hocks their possessions or turns to prostitution to support a cannabis habit.

The low rate of marijuana-related harm is certainly encouraging. Nevertheless, antiprohibitionists argue that it may not stem from the drug laws. Legal sanctions against intoxicants can decrease their use in many ways. Laws may increase fear of arrest, decrease the availability of drugs, or raise prices. Marijuana laws, however, do not appear to have as much impact in these domains as prohibitionists might hope. Most people who do not use marijuana claim that they abstain because they have no interest in the drug; they do not report that a fear of legal problems motivates their behavior (Maloff, 1981). Fear of arrest is actually remarkably low. People who use cannabis but do not sell it probably only have a 2% chance of arrest per year (MacCoun, 1993).

Few people claim that they would change the amount they used if marijuana were legalized (Johnston, Bachman, & O'Malley, 1981). A poll of 1,400 adults found that over 80% claimed that they would not try the drug even if it were legal (Dennis, 1990). In fact, some users joke that the drug would no longer produce intoxication if it were legalized (Lenson, 1995). These data require cautious interpretation. People are notoriously poor at explaining why they behave in certain ways, or how they would act if conditions were dramatically different. A long period of legalization may alter these attitudes dramatically, making people more likely to try the drug if sanctions disappeared. Nevertheless, few report that fear of arrest changes their marijuana consumption.

The impact of marijuana laws on availability of the drug also appears small. Every year since 1975, over 80% of high school seniors have reported that marijuana is fairly easy or very easy to purchase (Johnston, Bachman, & O'Malley, 1996). Most teens find beer more difficult to buy than cannabis (Center on Addiction and Substance Abuse [CASA], 1996). The price of marijuana undoubtedly increases because of its illegal status. Yet users may be relatively insensitive to price, at least to the extent that it can be manipulated by legal sanctions (MacCoun, 1993). The drug is actually quite cheap compared to other intoxicants, costing a couple of dollars per hour of altered consciousness. Marijuana intoxication may be less expensive than seeing a movie in a theater. Thus, these data suggest that the current laws may have little impact on use because they fail to create fear of legal sanctions, decrease availability, or raise prices enough to eliminate demand. Nevertheless, studies of areas that have actually changed their laws may provide a better picture of the impact of different policies.

The Decriminalization Experience

Areas where marijuana is decriminalized can reveal some of the potential effect of legal sanctions. Findings from these areas, however, conflict. Some studies reveal little change in use; some suggest increases after many years of relaxed policies. The Netherlands, Australia, Italy, and

Spain have removed criminal sanctions for possession of a few grams of marijuana.

The decriminalization of cannabis and hashish in the Netherlands remains widely misunderstood. Marijuana and associated products remain illegal as part of international treaty. In 1976, the Dutch decided to eliminate enforcement for violations involving sale or possession of up to 30 grams. Thus, police do not enforce laws against marijuana if the amount involved is too small. Dutch policy makers hoped that this change in enforcement might help separate the marijuana market from the sale of drugs with markedly worse negative consequences. After this policy began, many coffee shops began selling marijuana and hashish. Legal guidelines for these cafes developed. They may not advertise, permit gambling, sell hard drugs or alcohol, admit anyone under 18, or operate near a school. International pressure forced a reduction in the maximum amount of an individual sale to 5 grams in 1995. Nevertheless, patrons can buy from six different shops and obtain 30 grams quite easily.

These controversial policies attracted considerable attention. Critics of Dutch decriminalization predicted that drug use would skyrocket. Yet marijuana consumption in the Netherlands remains comparable to use in the United States. Critics were particularly concerned about undermining the prevention of drug use among youngsters. Some studies suggest that rates of use are actually lower in the Netherlands than in areas where harsher penalties continue. Recent data reveal that only 21% of Dutch citizens age 12–18 ever tried the drug, compared to 38% of Americans that age. Recent studies found that only 11% of Dutch youth reported using marijuana in the past month, but 18% of Americans the same age smoked cannabis in the previous four weeks.

Despite these data, marijuana use may have increased for one age group since the policies changed. The rates of use within the Netherlands have increased for the specific subset of the population age 18–20. In 1984, 15% of 18–20-year-olds had tried the drug; by 1996, the percentage increased to 44% (de Zwart, Stam, & Kuiplers, 1997; NIDA, 1997). An increase 8 years after the initial steps of decriminalization may reflect a gradual change in attitudes about the drug. The fact that the increase appears specifically around the age of permitted use may show some sort of rite of passage into adulthood. Perhaps turning 18 years old leads to a party devoted to trying the drug, much the way Americans get drunk when they turn 21.

Many people fear that decriminalization of marijuana creates higher rates of use of other drugs. Data from the Netherlands do not support this concern. Although comparisons between different countries remain difficult to interpret, heroin and cocaine use remains lower in this area than in countries with harsher penalties for marijuana. The number of heroin users per capita in the United States (308 per 100,000 residents) dramatically exceeds the number in the Netherlands (160 per 100,000 residents) (Dutch Ministry of Health, Welfare, and Sport, 1995). In addition, fewer teens try cocaine in the Netherlands than in the United States (Zimmer & Morgan, 1997). Perhaps the decriminalization of marijuana has weakened its connection with these other substances. Given that people can obtain cannabis in shops where cocaine and heroin are forbidden, purchasing marijuana no longer must lead to exposure to harder drugs. Increased availability of cannabis may have decreased interest in other intoxicants, too.

Not all data on decriminalization polices come from the Netherlands. Two of Australia's eight territories also decriminalized possession of less than 25 grams of marijuana. Consumption of cannabis in a public place and sales of the drug remain illegal. People in South Australia and Australian Capital Territory face fines up to $150 for possession. Offenders receive a Cannabis Expiation Notice, much like a traffic ticket, and must pay their penalties within 60 days. They also have their cannabis confiscated. Law enforcement officers find these notices easier to issue and sustain than a full arrest. Thus, the number of offenses has increased dramatically. In a sense, this approach has increased the likelihood of penalties despite decreasing their severity. Yet rates of marijuana consumption in the decriminalized areas remain comparable to the rates in Australian territories with harsher penalties (Ali et al., 1998; McGeorge & Aitken, 1997; National Drug Strategy, 1995).

Although data from the Netherlands and Australia suggest that decriminalization may not increase marijuana consumption, many Americans see the experience of other countries as irrelevant to the United States. Comparing data across different countries with different policies creates many interpretive problems (MacCoun & Reuter, 1997). Thus, perhaps only data from within the U.S. borders remain relevant. Eleven states essentially removed criminal penalties for possessing small amounts of marijuana by 1979: Alaska, California, Colorado, Maine, Minnesota, Mississippi, Nebraska, New York, North Carolina, Ohio, and Oregon. Decriminalization in America has led to little change in marijuana use, much like the experience in the Netherlands and Australia. Use by high school seniors in decriminalizing states did not differ from use in other states where sanctions remained (Johnston et al. 1981). Oregon, Maine, and California showed little change in use by adults after decriminalization (Maloff, 1981). Other states may have had comparable experiences.

Prohibitionists emphasize that federal laws against marijuana remained throughout these periods of state decriminalization. Federal laws may have helped keep cannabis consumption from skyrocketing in the decriminalized states. Fans of decriminalization point to data from these states and the Netherlands and Australia to suggest that harsh criminal penalties for marijuana possession may not deter use any more than simple, civil fines. Nevertheless, prohibitionists suggest that decriminalization will undermine the perceived harmfulness of the drug, leading to increased use and problems many years in the future. A change in federal laws may have long-term implications that data from other countries or a few states cannot reveal.

Estimating the Costs of Marijuana Prohibition

Whatever the benefits of marijuana prohibition, the laws also generate costs. These include the price of law enforcement and incarceration. In addition, the taxes that a legal marijuana market could generate are also lost. Other costs may transcend finances. Current methods of enforcement may lead to a loss of civil rights and decreased respect for the law. These costs all prove difficult to estimate. Law enforcement officials do not break down their expenses by the type of drug they anticipate eliminating. Budget information can help the estimation process. The federal government spends $15.7 billion annually on drug prohibition (Office of the National Drug Control Policy [ONDCP], 1997a). State and local governments spend approximately $16 billion annually enforcing drug laws, for a total of nearly $32 billion (ONDCP, 1997b). Approximately 43% (642,000) of the 1.5 million drug arrests in 1996 were for marijuana offenses (FBI, 1997). If all arrests were equally costly, America spent $13,760,000,000 on marijuana arrests—approximately $21,400 for each one. Some arrests undoubtedly cost more than others. Even if marijuana enforcement cost only half this amount, Americans have clearly spent billions in an attempt to eradicate this drug and will likely continue to do so.

Another potential financial cost of marijuana prohibition concerns tax revenues lost to the underground market. Proponents of legalization emphasize that taxing marijuana could fund drug prevention or treatment programs or help pay the national debt. Moralists in support of prohibition find this argument reprehensible. Nevertheless, if the drug were legal, it would become markedly cheaper to produce. The expense associated with hiding the crops from law enforcement and poachers would likely decrease. (As a comparison, the price of alcohol and its production dropped after the repeal of the Eighteenth Amendment.)

Taxes on marijuana could maintain its current price. The tax might help pay to ensure a quality product, labeled according to potency, and free from contaminants like pesticides. Ideally, this arrangement would encourage users to purchase taxed, legal cannabis rather than marijuana with unknown characteristics from the underground, illicit market. The exact amount lost under prohibition is difficult to estimate, but probably approaches several billion dollars. Current estimates suggest that marijuana is the fourth most valuable crop in the United States, behind corn, soybeans, and hay (NORML, 1996b). This estimate assumes a wholesale price near $2,700 per pound, generating approximately $15 billion. Retail prices for sales of markedly less than a pound can run three times that amount or more. Taxes and licensing fees associated with this market might generate several billion dollars each year (Kleiman, 1992). Estimates that include revenue from income taxes for workers in the marijuana industry suggest prohibition may cost over $10 billion per year in lost funds (Rosenthal & Kubby, 1996). This money might finance programs to discourage problematic use of the drug.

Other losses attributed to prohibition include concerns about civil rights and respect for the law. The Fourth Amendment of the Bill of Rights protects citizens against searches that lack probable cause. Recent drug cases have permitted law enforcement officers considerable latitude in their interpretation of reasonable suspicion and the appropriate justification for arrest. Some judges assert that this latitude is necessary given the creative strategies that drug traffickers employ. Stricter guidelines might limit the number of arrests. Critics argue that this unrestrained approach essentially violates the Fourth Amendment. They also detail misunderstandings that have led to illegal searches and even loss of life because activities in poor neighborhoods were misinterpreted as drug deals. Putting an exact price on these drawbacks of prohibition is impossible. Balancing them against lost arrests also defies simple analysis (Ostrowski, 1998).

These concerns about violations of civil rights associated with law enforcement may contribute to a general disrespect for the legal system. Many citizens view penalties for possession of marijuana as too harsh. They also see minorities as unfairly targeted for arrest. Data support this assertion. Caucasians are underrepresented in marijuana arrests. Given the number of Caucasians who report using marijuana, a disproportionate number of minorities find themselves in court for marijuana crimes (Mandel, 1988). Many people also consider marijuana prohibition hypocritical given the legal status of alcohol and tobacco. These factors undermine the efficacy of drug regulations and may cause some law-abiding citizens to lose respect for legislation in general (Packer, 1968).

Alternative Plans

Many people argue for the status quo on cannabis policy, but a number of authors suggest alternatives. The range of proposals for changes in marijuana legislation encompasses everything from harsher penalties to completely unregulated legalization. Each relies on moral arguments or predictions about potential costs and benefits of new policy. Most fall between extremes, focusing on decriminalization or legalization with increased regulation. The regulations frequently include taxes, controlled sources of sales, strict age limits, penalties for inappropriate use, and special licenses for users. All these proposals have advantages and disadvantages, much like current policy. Some reformers recommend federal changes that should apply throughout the country; others assert that marijuana's regulation should parallel alcohol's and become the authority of each individual state (Benjamin & Miller, 1991).

Getting Tough

Efforts to increase penalties generally rely on their ability to deter any behavior. Proponents of stiffer sanctions assert that they would eliminate drug use. Proposals include a very extreme approach typified by the expression "If you try, you die." In 1990, when he was chief of police in Los Angeles, Daryl Gates suggested that occasional marijuana smokers should be "taken out and shot" (Gordon, 1994). This approach to casual consumption would likely inhibit many users. Malaysia and Singapore impose the death penalty on drug trafficking. Casual users in these countries can receive imprisonment in centers purportedly designed for rehabilitation. These centers include such therapeutic activities as solitary confinement and hard labor. Both countries have arrest rates for drug possession that are 30% lower than those in the United States (Benjamin & Miller, 1991). For obvious reasons, reports of use are probably extremely low.

Critics of these policies bemoan the violations of civil rights that their enforcement would likely require. They also point to data that suggest that the severity of penalties may have little impact without increases in the probability of arrest (MacCoun, 1993). Appreciable increases in the probability of arrest could prove extremely expensive. More law enforcement personnel, court costs, and prison facilities could cost as much as $150 billion. In addition, many users may turn to alcohol or prescription drugs in an effort to substitute for marijuana. The health impact of these changes could prove quite expensive because the negative consequences associated with problem use of these drugs may be more severe than those related to cannabis. Moralists in support of prohibition, particularly those who appreciate these policies, suggest that the decrease in drug use is worth the price of increased enforcement.

Decriminalization

Decriminalization proposals generally range from those employed in a few states in the United States to those closer to situations in the Netherlands. Several states in America keep production and distribution of marijuana illegal, but remove the risk of arrest for possession of small amounts for personal use. Civil penalties remain, with fines up to about $500. This approach minimizes one of the most dramatic negative consequences of marijuana consumption: jail time. Data suggest that law enforcement expenses also decline after decriminalization (Aldrich & Mikuriya, 1988). As previously mentioned, rates of use have not skyrocketed relative to areas that maintain criminal penalties. Some increased reports of use may arise from a growing willingness to admit to smoking marijuana after criminal penalties disappear.

Despite the advantage of money saved with little increase in consumption, disadvantages to decriminalization remain. This approach does little to combat the underground market. Exposure to this underground market also may increase contact with drugs markedly more harmful than marijuana. The potential tax dollars lost to prohibition continue under decriminalization. A decriminalization format closer to Dutch policy may combat some of these problems but may also create new ones. Removing penalties and adding taxes for sale of small amounts, if properly regulated, could increase revenues. This approach has not increased use in the Netherlands; perhaps U.S. residents may respond comparably. Assuming that regulation's price did not exceed the taxes generated, this approach could earn money and minimize the underground market (Kleiman, 1992). Nevertheless, rates of use may creep upward over many years in ways that data currently available cannot reveal. Problems would likely increase as use increased, with unknown associated expenses.

Legalization

Legalization proposals are dramatically different from any plans for decriminalization. They also show considerable range. A libertarian, free-market proposal suggests regulations of marijuana should be similar to those for any other goods for sale. These plans often rest on moral arguments for drugs as property (Szasz, 1992). Some legalization proposals employ the "tomato model," suggesting rules for cannabis should parallel those for standard agricultural products (Evans, 1998). These approaches would certainly eliminate expenses related to law enforcement and any concerns about violations of civil rights, but the impact on use and problems proves difficult to estimate. No country or state has actually legalized in this fashion; estimating the associated changes in use and problems proves extremely difficult. As mentioned earlier, many who do not use cannabis claim legal sanctions have little to do with their decision to abstain. Nevertheless, a long era of unregulated legalization may change these reports. The ingenuity of advertisers and marketers might lead to considerable increases in use.

Other proposals generally include markedly more regulation. These regulations were designed to limit use, particularly by minors. Most propose taxes that would fund law enforcement, drug abuse prevention, or local schools. Various bills have appeared in state legislatures over the last 30 years. For example, in 1971, the New York Senate reviewed a bill that would make marijuana control comparable to alcohol's, with additional limits on advertising. The Cannabis Revenue and Education

Act appeared in Massachusetts in 1981 and proposed taxed, regulated production and distribution of cannabis, with half the proceeds funding public education against marijuana abuse.

The Cannabis Revenue Act, a federal bill proposed in 1982, offered different options to states. All states could, of course, continue prohibition if they desired. They could allow federally licensed retailers to sell one-ounce packages bearing cautionary labels and a tax stamp. The Act also permitted states to form alternative plans. Sales to minors and driving under the influence would continue to carry severe punishments. Obviously, this Act did not receive enough support to become federal law. Oregon and Pennsylvania state legislatures reviewed comparable bills the following year. The bills in both states limited sales to state-run stores. These reforms have yet to pass any legislature, so their impact remains unknown. Nevertheless, they have the potential to drop law enforcement costs and minimize the underground market (Evans, 1998). They could also inadvertently increase consumption and problems.

Other reformers suggest additional regulations, including licensing users and limiting the amount they may purchase. This approach would provide a specified supply of marijuana to licensed users, perhaps through mail-order or state-run stores (Kleiman, 1992; Nadelmann, 1992). This strategy has the advantages of other legalization proposals. The government could tax the drug. Most consumers would prefer legal cannabis with competitive pricing, detailed labeling, and assurance of quality, to the underground market's product. This approach would minimize the exposure to other drugs that often accompanies purchases in the underground market. The expenses related to law enforcement and various violations of civil rights would decrease dramatically. Each purchase could also include a detailed pamphlet explaining how to prevent harm from the drug, including information on self-help groups or treatment centers.

People experiencing problems with the drug might choose to turn in their licenses. Obtaining the license might require an exam comparable to the written driver's test. To obtain a license, users would have to demonstrate a thorough understanding of relevant laws, as well as recommendations for preventing problem use. Any inappropriate behaviors, including driving under the influence, public intoxication, convictions for other illicit drug use, or providing the drug to unlicensed users, could lead to loss of the license in addition to other penalties. This approach could help minimize irresponsible behaviors under the influence, as well as distribution to minors or people with drug problems.

Specifying a limited amount for purchase remains difficult. Smaller quantities may not only decrease the probability of problem use but also may keep the underground market alive. Taxing marijuana at a rate that would make the legal price appreciably less than the underground price might help eliminate illicit sources of the drug. Unfortunately, this arrangement may encourage licensed users to sell their legally obtained cannabis at a profit—an offense that would lead to penalties and a loss of the license.

Estimating a reasonable limit requires a number of assumptions. Individual differences in reactions to the drug are quite large. The average amount of time intoxicated per unit dose would have to be determined for the marijuana for sale. If one gram led to roughly 4 hours of intoxication, 52 grams per year (a little less than two ounces) would permit licensed users to spend roughly 2% of their time under the influence of the drug. This 4 hours per week would be more time than most people spend in houses of worship, and roughly one-seventh the amount of time the average person spends watching television (A. C. Nielson Co., 1998). Those who use less of the drug need not purchase as much. Those who require more for medical reasons could arrange prescriptions for larger amounts.

Salient drawbacks to the licensing proposal concern the inestimable impact it may have on the number of marijuana smokers, and the ethical issues involved with controlling access to the names of licensed users. Getting a license could turn into a rite of passage into adulthood, which might increase interest in consumption. Yet licensing users and limiting purchases could minimize negative consequences of the drug, even if more people decided to consume it. The privacy of an individual's license remains a complex issue. Although data are unclear about the health risks of moderate use of marijuana, insurance companies may want to alter the premiums of licensed users.

Additional questions about revealing the status of individual licenses relate to employment. Although marijuana intoxication may enhance performance in some dull, repetitive jobs (Carter, 1980), it may impair performance on other important ones. Some proposals suggest that licensed users may not qualify for certain jobs, such as air-traffic controller. These proposals may confuse licensing use during free time with intoxication during work hours. Just as millions of workers who purchase alcohol legally do not attend their jobs drunk, a licensed marijuana user need not report to work high. Yet the fear of violations remains too great. These general fears inspired dramatic efforts related to drug testing in the workplace, another complex legal issue.

Drug Testing at the Workplace

Employers have an understandable interest in maximizing the performance of employees. Many are concerned that drug users may work less efficiently, have more accidents, or use more medical benefits. Other employers may favor hiring abstainers because of moral objections to drug use. Over 80% of major U.S. firms test for drugs, spending millions of dollars in the process. Opponents of employee drug testing view it as a degrading experience that qualifies as an illegal search and seizure. They also argue from a more utilitarian perspective that the costs of testing outweigh the benefits.

Studies of cannabis users in the workplace may help answer some relevant questions about efficiency, accidents, and medical benefits. Few data address the efficiency of marijuana smokers at their jobs. One study reports they earn higher wages (Kaestner, 1991). Intoxication on the job would likely impair performance, though some workers report improved manual labor after smoking (Carter, 1980). In fact, very few people actually use illicit drugs at work. Any residual effects from marijuana consumption off-duty appear slight to nonexistent (Normand, Lempert, & O'Brien, 1994). A study of accidents in post office employees found no differences between drug users and nonusers (Zwerling, Ryan, & Orav, 1990). This absence of an effect likely stems from the infrequency of drug consumption at work. Marijuana appears to have no impact on health benefits, either. A study of a large group of HMO patients in California's Kaiser Permanente program compared medical costs of users and nonusers and found no differences (Polen, 1993).

Given these limited benefits of identifying marijuana smokers, the costs of drug testing may only appear worthwhile to those with a strong moral opposition to cannabis. A study of the federal government's $11.7 million drug testing program examined the efficiency of the procedure. Given the large number of abstainers and the price of the tests, identifying a single drug user cost $77,000. Proponents of the program argue that the tests deter drug use in employees, but the rate of positive tests parallels the reported rates of drug use in the nation. Thus, government workers use illicit drugs at the same rate as others, suggesting that the tests do not deter consumption.

In addition to their direct cost, drug tests decrease productivity because employees are not working while providing hair or urine samples. Data suggest that computer firms with drug testing programs actually score lower on productivity measures than comparable firms that do not test for drugs. The impact on employee morale can also be particularly negative. Some companies have dropped preemployment drug testing because it impaired their ability to hire qualified applicants. Alternatives to drug testing include many money-saving strategies that lack the degradation often associated with drug tests. Most approaches focus on employee performance rather than drug consumption. Jobholders whose work needs improvement receive appropriate feedback and employee assistance. Individuals in positions that require optimum performance to ensure safety can complete brief cognitive tests prior to the beginning of work. Supervisors can send impaired workers home whether their deficits stem from intoxication, fatigue, or illness (Maltby, 1999).

Many people dislike drug testing in the workplace and take extreme steps to undermine its efficacy. A small industry has developed in reaction to widespread drug testing. This industry sells products designed to enhance the chances of testing negative despite drug use. A number of shampoos purportedly mask drug use for the hair test. Data suggest that they may decrease concentrations of cannabis metabolites in hair, but a single administration will not bring them to undetectable levels (Rohrich, Zorntlein, Potsch, Skopp, & Becker, 2000). Several compounds added to urine may create false negatives, but laboratories now test for them. Drug lore suggests false negatives increase with the ingestion of various herbs, cranberry juice, vinegar, mineral oil, lemon juice, or diuretics. These approaches also have no empirical support, except for effusive urban legends (Coombs & West, 1991).

A few legends about drug testing are based in fact. Drinking huge amounts of water likely lowers the concentration of metabolites in the urine. The first urine of the day may contain more metabolites than those given at other times. Scheduling tests for later in the day may increase the chances of false negatives. Otherwise, long periods of abstinence are the only guarantee of a negative drug test. The numerous products and extensive clinical lore designed to combat drug testing reveals a negative attitude toward this indirect side effect of current attitudes and policies.

Conclusions

Despite cannabis's long history, American federal laws about the drug did not appear until the Marijuana Tax Act of 1937. Stories of drug-induced mayhem led to this first legislation. Attitudes about cannabis have shown considerable variation over the 60 years that followed. Penalties initially grew more harsh. Selling the drug to a minor could bring the death penalty in some places during some eras. In the 1960s, as more middle- and upper-middle-class young people experimented with the drug, attitudes changed. By the end of the 1970s, eleven states had decriminalized possession of the drug for personal use. Owning less than an ounce for individual consumption could lead to little more than a fine. The 1980s and 1990s have seen not only a return to criminalization in some states but also a new movement for legalizing cannabis for medical purposes.

Debates related to marijuana prohibition claim to focus on moral and pragmatic issues. Moral arguments in support of prohibition generally portray the drug as unethical or corrupt. Pragmatic arguments in support of prohibition rest on concerns about health consequences, use by youth, and the transition to other illicit drugs with worse effects. Moralists against prohibition assert that current laws violate rights guaranteed by the Constitution, including freedom of religion, privacy, and property. Pragmatic arguments against prohibition focus on the price of enforcing the laws relative to the harm the drug actually causes.

Alternatives to the current laws remain numerous and varied. A few prohibitionists assert that extreme penalties rigorously enforced would decrease use dramatically, but the expense of such a program and the associated potential for violations of civil rights make it unfeasible. Evidence from countries with stiffer penalties, such as Singapore and Malaysia, suggest that even these efforts would not eliminate the drug. Proponents of decriminalization report that minimizing penalties has saved many areas considerable expense with little increase in consumption. Nevertheless, decriminalizing the drug does little to combat the underground market. These underground sales command large sums of untaxed revenue and may expose users to drugs with markedly worse negative consequences.

Some proponents of legalization suggest innovative strategies that could have all the benefits of decriminalization with the added advantage of altering the underground market. Government programs that would license marijuana users and limit the amount they may purchase could provide tax revenue and avoid exposing users to other drugs. These strategies would also save large sums currently devoted to law enforcement. Yet recent attempts to change laws in this direction have met with considerable resistance. After a long period of legal use and more than 60 years of changing legislation against the drug, future trends seem difficult to predict. Given the moral and pragmatic arguments given by both sides, it seems clear that this is one area where more research will not have as much impact as changes in political climate.

11
Treatment for Marijuana Problems

This chapter addresses ways to alleviate marijuana problems. Some of the intricacies of treatment research appear first, followed by a review of studies focused specifically on therapy for troubles that stem from smoking too much cannabis. Suggestions for ways to improve therapy follow, including descriptions of three promising treatments for substance abuse. These treatments include cognitive-behavioral therapy, twelve-step facilitation, and motivational interviewing. This chapter is not a substitute for substance abuse treatment but may serve as a guide to the approaches available for limiting marijuana-induced harm.

The path from marijuana use to marijuana problems varies from person to person (Newcomb & Earleywine, 1996); the path away from these problems varies, too. Estimates suggest that 9% to 15% of cannabis users develop some problems with the drug (NIDA, 1991; Weller & Halikas, 1980). Some folks quit without formal interventions; some quit through treatment; and some never quit. Contrary to popular belief, the majority of people who abuse drugs stop on their own. Most decrease their use without the help of drugs, books, organizations, clinicians, or coaches (Peele, 1998; Vaillant, 1983). They often report that they quit in an effort to avoid problems. Some mention that family support, a new job, or a sense of accomplishment may have assisted their change.

Many abstinent people report a critical incident that inspired them to quit. A cannabis smoker might stop after saying something embarrassing while high, developing a bad cough, or getting arrested for possession. Research suggests spontaneous quitters may not have more of these events in their lives than people who keep using. The events do not have to be particularly dramatic. Some users abandon cannabis after developing new hobbies or becoming parents. No clear set of circumstances or incidents leads all people to reduce drug use. Many researchers assert that spontaneous quitters were never addicted in the first place. Others suggest that intensive study of spontaneous quitters can reveal a great deal about the addictive process (Sobell, Sobell, Cunningham, & Toneatto, 1993). In addition to those who quit on their own, some people never quit. They may experience problems that are not particularly debilitating, find a level of functioning that suits them, and continue using. The number of these functional, problem marijuana users remains unknown.

In addition to those who quit on their own or never quit, some people turn to professionals for help. These may be the people who have a particularly difficult time quitting on their own. Therapies for the addictions do not appear completely effective. Treatment studies reveal that more than half of users of any drug who successfully quit eventually use again (Brown, 1993). Psychological treatments for other problems are more successful. For example, people with panic disorder (Bruce, Spiegel, & Hegel, 1999), anxiety (Yonkers, Warshaw, Massion, & Keller, 1996), and depression (Evans et al., 1992) show much better responses to therapy. Drug treatment programs have considerable room for progress. Perhaps they try to do too much for too many people.

Research on treatment may not reveal much about the way therapies proceed outside the research settings. For example, most drug programs must treat people with any sort of chemical dependence. Urban crack abusers may find themselves with rural alcoholics in the same facility. These people may experience different troubles that require markedly different interventions. In contrast, some research studies concentrate on participants experiencing problems with a single drug. This single-drug approach may prove more effective for marijuana problems. Treatments specifically designed for difficulties related to marijuana may attract more problem smokers than programs designed for addictions in general (Roffman & Barnhart, 1987). Perhaps problem marijuana users do not see themselves as comparable to people experiencing troubles with other drugs. Perhaps they are concerned that other participants addicted to crack cocaine, heroin, or alcohol might not take problems related to marijuana very seriously.

Research on Treatment for Cannabis Problems

One of the few empirical studies devoted exclusively to treatment for marijuana problems helped over a one-third of the smokers eliminate negative consequences completely, and 14% maintained abstinence for a year. The study compared relapse prevention techniques to a social support group (Stephens, Roffman, & Simpson, 1994). Relapse prevention techniques stem from the cognitive-behavioral model of treatment, and social support may play a role in twelve-step approaches. The details of these strategies appear at the end of this chapter. The researchers identified 212 people (161 men and 51 women) who were appropriate for the study. Most were employed, educated, Caucasians in their early 30s. On average, they had smoked marijuana for 15 years. They used on an average of 81 days in the previous 90. Some smoked 4 or more times in a day. They showed no difficulties with drugs other than cannabis. Their concerns included trouble decreasing their use, negative feelings about smoking marijuana, procrastination, decreased self-confidence, memory loss, and withdrawal symptoms. Many also reported experiencing financial difficulties and complaints from their loved ones (Stephens, Roffman, & Simpson, 1993). This sample clearly qualified as heavy users experiencing adverse effects.

The two treatments emphasized different aspects of cannabis use. The relapse prevention treatment focused on identifying feelings, thoughts, and situations that might increase the chance of smoking marijuana, and planning alternative actions that did not include the drug. The social support group concentrated on identifying meaningful others who could assist in maintaining abstinence, particularly during difficult times. Despite these different emphases, the two treatments shared many factors. The goal for both therapies was abstinence. Both used a group format. Twelve to 15 participants met with two therapists for two hours per session. Therapists scheduled 10 meetings in the first 3 months; participants attended an average of 7 or 8 of them. Therapists held two additional sessions 3 and 6 months after treatment ended. These booster sessions reviewed the previous material. Researchers often schedule these booster sessions in an effort to improve and assess long-term outcomes.

Participants who attended these final sessions provided urine samples and answered written questions about marijuana use and problems. Urine screens revealed that the written answers were very accurate. The clients also answered questionnaires mailed to their homes at 1, 9, and 12 months after treatment. In addition, a friend or relative of each participant reported on his or her use. These collateral reports agreed with the participant information very often, helping to confirm accuracy. Thus, the clients provided multiple measures of marijuana use and problems at 1, 3, 6, 9, and 12 months after treatment ended.

Clients improved with treatment. The relapse prevention and social support therapies were equally effective at minimizing marijuana problems. One difference between the two approaches did appear: the relapse prevention group smoked marijuana fewer times per day than the social support group. Otherwise, the groups were equal in their numbers of problems and the number of days that they smoked. Over 60% of the participants remained abstinent for 2 weeks after a specified quit date. Nevertheless, only 14% actually maintained abstinence for the entire year after treatment. Some participants eliminated problems, even though they did not remain abstinent. People who reported no marijuana problems, and who cut the number of days they got high in half, were considered improved. Using these criteria, 36% of the sample appeared improved one year after treatment, 31% had been abstinent or improved for that entire year. Despite the decreases in marijuana consumption and problems, the sample reported drinking alcohol more often. They also experienced a few more alcohol problems after treatment. Perhaps some turned to alcohol after limiting their cannabis use.

An important caveat about these results concerns participants who dropped out. The researchers only described outcomes for the 167 people who recorded their progress at all 5 of the follow-up points. We cannot know the status of the other 45 users who did not complete these measures. This level of attrition (21%) is typical of treatment studies. These lost participants may have made as much progress as they wanted and decided to stop their involvement in the study. Maybe they were too busy to fill out questionnaires and pee in cups. Perhaps they made little progress and quit treatment. One way to gain some insights into the status of people who left the study requires comparing them to the ones who remained. If these two groups did not differ prior to the study, perhaps their success with treatment was also comparable.

In fact, the 45 people who did not complete all the assessments differed significantly from the other participants. They reported more marijuana problems initially and had longer histories of use. A longer, more problematic involvement with cannabis apparently increases the likelihood of missing follow-up assessments. Some might argue that these people probably had worse outcomes than those who completed follow-ups, given that they were experiencing more problems initially. Data from this same research group suggest that those who did not complete the study were younger, earned less money, and reported more psychological distress (Roffman, Klepsch, Wertz, Simpson, & Stephens, 1993). Thus, the reported success rates for those who completed all the assessments may overestimate the success of the treatment for the lost participants.

Interpretations of this treatment study depend on a frame of reference. The idea that almost two-thirds of those in treatment could not eliminate their marijuana problems and over 85% continued to use the drug might seem unimpressive. These numbers look particularly discouraging in light of the 45 people who did not complete the outcome measures. Nevertheless, these data do suggest that problem users are not doomed to a lifetime of negative consequences. The sample obviously does not include those who were able to quit on their own. (On the other hand, it also does not include problem users with no desire to quit.) But for people who seek treatment and remain in therapy, the 36% improvement rate serves as our best estimate of the efficacy of these two approaches. These results parallel those from studies of people dependent on alcohol, cigarettes, and heroin (Hunt, Barnett, & Branch, 1971). Given the success rate of therapies for other addictions, both relapse prevention and social support appear equally encouraging for decreasing problematic marijuana use. In addition, these therapies only required a dozen sessions in a group format, suggesting minimal cost. Perhaps simple, cost-effective improvements might enhance the outcomes for these treatments.

Potential Improvements for Treatment of Cannabis Problems

Therapists and researchers have suggested several changes that might improve treatments for marijuana problems. These include using individual therapy, adding an inpatient stay in the hospital, increasing the number of sessions, and providing medications to help abstinence. Data suggest that some of these changes may provide more help than others. Many drug programs have focused on individual rather than group treatment. At first glance, the group format may seem less promising. Time spent one-on-one with the mental health provider might provide extra attention, specific interventions, and a different therapeutic relationship.

Although no data address the issue of group versus individual treatment for marijuana problems, a study of cocaine abusers actually found some advantages for group treatment (Schmitz et al., 1997). This study revealed fewer cocaine-related problems for participants in group therapy compared to those in individual. Perhaps group treatments provide social support that enhances outcome. Thus, individual sessions may not be essential for improving therapy for marijuana problems.

Another popular approach to drug treatments includes inpatient, hospital stays. Hospital patients may have less access to illicit drugs, improving their chances of maintaining abstinence. Yet a review of many studies suggests that staying in a hospital instead of using an outpatient program may not prove worth the cost (Miller & Hester, 1986). Other studies support inpatient treatment for a subset of drug abusers (Moos, King, & Patterson, 1996). No data address the efficacy of hospital stays for people experiencing difficulties from marijuana. Few problem users of cannabis express much desire to enter the hospital for treatment. This approach may offer little improvement in outcomes if no one is willing to use it.

Longer treatments also have an intuitive appeal. More sessions might give extra opportunity to learn skills and discuss problems. Increased counseling has created better outcomes for many people with drug problems (Fiorentine & Anglin, 1997). Again, no data address marijuana troubles directly, but studies of other drug problems may apply. An examination of a therapeutic community of drug abusers found a 6-month program led to better results than a 1-month treatment (Bleiberg, Devlin, Croan, & Briscoe, 1994). A study of over 2,000 drug users revealed longer treatments could lead to less use of heroin and cocaine (Hser, Grella, Chou, & Anglin, 1998). Longer attendance in an aftercare program led to better outcomes for alcoholics, too (Trent, 1998). Yet long treatments are clearly not essential. Some people decrease negative consequences of drugs with brief interventions (Tucker, Donovan, & Marlatt, 1999). Perhaps longer treatments could improve outcomes for people experiencing marijuana problems who do not respond to brief therapies. People with more severe problems may need continued support to maintain abstinence.

Pharmacological interventions in combination with psychological treatment also have many supporters. Although some argue that pharmacological treatments for the addictions simply substitute one drug for another (Cornish, McNicholas, & O'Brien, 1995), this approach is consistent with the biopsychosocial and medical models of addiction. Medications designed to treat substance abuse usually function in one of four ways: (1) they can provide the drug in a less harmful way; (2) they can react adversely when combined with the drug of abuse; (3) they can lower the craving and euphoric drug effects; or (4) they can combat the symptoms that may have initially inspired use of the abusable drug. Applications of each approach to marijuana problems have been limited so far.

Some pharmacological interventions provide small doses of a drug to decrease craving and ease withdrawal. The nicotine replacement therapies work this way. Nicotine gum, patches, and nasal sprays may help cigarette smokers quit if combined with behavioral techniques (Fiore, Smith, Jorenby, & Baker, 1994; Sutherland et al., 1992). This approach has never been attempted in the treatment of marijuana problems. Although alternative ways to administer THC exist, including orally administered dronabinol and the marijuana suppository, replacement therapies may have too high an abuse potential to treat negative consequences related to cannabis use. This approach does not appear popular in any treatment community.

Some medications cause an adverse reaction when combined with an abusable drug, which may assist recovery. For example, disulfiram (Antabuse) makes people who drink alcohol feel ill, which may help alcoholics avoid relapse (Adelman & Weiss, 1989). Silver acetate makes cigarette smoke produce a nasty, metallic taste, which may assist smoking cessation (Hymowitz, Feuerman, Hollander, & Frances, 1993). Nevertheless, no one has identified a drug that would consistently make marijuana aversive. Thus, this approach may not improve outcomes for problem marijuana users.

Other medications may block the rewarding effects of abusable drugs and help minimize craving. Naltrexone may decrease craving for opiates and alcohol, as well as lower their euphoric effects if they are consumed. Naltrexone has improved relapse rates for alcoholics (Volpicelli, Alterman, Hayashida, & O'Brien, 1992) and may help opiate addicts maintain abstinence (Holloway, 1991). Methadone decreases opiate-induced euphoria and craving and can enhance recovery from heroin abuse (Callahan, 1980). A medication that would block marijuana craving and its euphoric effects may help minimize problems, but no one has identified such a substance. A few substances block the CB1 receptor, but their safety and utility in the treatment of cannabis abuse and dependence remains unknown.

The medication strategies discussed above confront drug problems directly. Another approach concerns providing medication for drug users who use marijuana to combat psychological problems. A medication that alleviates these symptoms might minimize the need for marijuana. Thus, medication could decrease marijuana consumption indirectly by eliminating the state that may have inspired the cannabis use in the first place. Some people may use marijuana in an attempt to combat symptoms of mood disorders or other psychological troubles (Burton, 1621; Grinspoon & Bakalar, 1997). Medications with fewer side effects and negative consequences might help these people leave marijuana behind. Psychological treatments that focus on these symptoms might also help people reduce their marijuana smoking. For example, cognitive-behavioral treatments can decrease depressive symptoms and eliminate cannabis that may have been used as an antidepressant.

Thus, outpatient relapse prevention and social support remain the best (and only) formally investigated therapies for marijuana problems. Because the success rates have been less than perfect, the therapies may improve with a few alterations. Individual treatment and hospital stays are probably not essential. Longer treatments may increase success rates. Pharmacological interventions for coexisting psychological problems like depression or anxiety might also help treatment for marijuana problems. These data and studies of treatments for alcohol problems also suggest that three psychological treatment options appear promising. Descriptions of each appear in more detail next.

Promising Psychological Treatments for Marijuana Problems

At least three different approaches have shown considerable promise in minimizing the negative consequences of drugs. These include cognitive-behavioral therapy, twelve-step facilitation, and motivational interviewing. Cognitive-behavioral treatment focuses on changing the thoughts and situations that previously led to drug use. The relapse prevention treatment in the study described earlier derives from cognitive-behavioral theory. Twelve-step facilitation employs specific techniques to help people make good use of twelve-step treatment. These programs rely on social support as one aspect of treatment, which may parallel the social support group in the study above. Motivational interviewing uses assessments and interpersonal interactions to enhance decisions to alter problem behaviors.

No single study of marijuana problems has compared the three approaches. Each treatment has its strengths. An enormous project that contrasted the outcome of these three treatments for alcohol-dependent people found that all three were comparably effective (Project MATCH, 1998). The treatments share several factors, which may help explain their similar outcomes. All emphasize the client's responsibility for change. All treat the problem drug use as a phenomenon independent of the individual's value as a person. All three stress regular attendance and active participation in treatment.

Descriptions of these therapies do not reveal all their nuances. Any attempt to reduce a treatment to a few pages of text invariably fails. Even the briefest interventions for substance abuse defy simple explanations. Academic descriptions of psychotherapy often miss its potential for intimate and curative interactions. Stereotypical depictions of the process often emphasize education, empathy, encouragement, and occasional insights. Ideally, these combine to alter actions, diminish problems, and increase happiness. The techniques and rationales of each of these treatments only provide a limited picture of the way they actually proceed.

Although treatments differ in their methods and strategies, most require a meaningful relationship with a therapist. Therapists often believe techniques create change, but the relationship may serve as an equally important contributor (Strupp, 1989). The idea that the relationship is more important than specific strategies may help explain some of the similar outcomes created by different therapies (Wampold et al., 1997). Perhaps disparate treatments create comparable results because all rely on a therapeutic relationship. Manualized treatments, which clearly delineate specific material for each session, still lead to different outcomes with different therapists. The therapeutic relationship may account for these differences. Yet this relationship does not mimic the friendship and mentoring common outside of therapy. Even treatments for relatively simple problems, like bedwetting or a fear of dogs, require considerable skill on the clinician's part. Data clearly support psychotherapy's efficacy, but the mechanisms that lead to success remain unclear. Therapy works; no one knows exactly why (Dawes, 1994). The descriptions below describe the rationales for each therapy, but the treatments may not succeed for the reasons posited. All may depend, at least in part, on the therapeutic relationship.

Cognitive-behavioral Therapy

Cognitive-behavioral therapy for substance abuse focuses on altering environments, thoughts, and actions associated with drugs. Different aspects of people's environments may trigger undesired, problematic consumption. These triggers include external and internal factors. External factors include any person, location, or object associated with drugs. A roach clip, rock song, or ashtray might easily trigger a desire for a drug. Internal factors include thoughts and feelings linked directly or indirectly to the drug. Some triggers are direct and some are indirect. Direct factors are close to drug use, like craving and urges. Indirect factors also increase the chance of drug use, but their import is less obvious. These include frustration, anger, or even delight. Cognitive-behavioral therapy suggests that users learn to take drugs in reaction to these triggers, much the way people learn any other behavior. Therefore, they can learn to engage in new behaviors instead of problematic drug use by altering environments, thoughts, and actions.

The idea that drug consumption is a learned behavior does not fit some people's experience. Many regular users report that drug consumption occurs automatically, without much thought or effort. In fact, triggers often occur before drug consumption, even when users remain unaware of them. Through consistent effort, client and therapist can work together to identify environmental factors that lead to drug use. Then the client can learn to engage in new behaviors in reaction to these factors. Detailed assessments of when an individual used drugs in the past can help reveal times that may prove particularly difficult in the future. This information serves as a first step toward minimizing the negative consequences associated with consumption (Beck, Wright, Newman, & Liese, 1993).

The situations that precede using drugs often appear bafflingly diverse. For example, an assessment might reveal dramatic cannabis consumption prior to a party, after conflict at work, and every Saturday. The commonalities among these situations may not look obvious. The cognitive-behavioral model suggests that thoughts about the situations may contribute more to using drugs than the circumstances themselves. Thus, each environment may elicit specific thoughts—often something like "marijuana is the only way to enhance this experience." These thoughts may prove easier to alter than the situations, so they become an essential focus of cognitive treatment.

The cognitive-behavioral model suggests that people carry a set of underlying beliefs with them. Certain situations activate these beliefs, eliciting specific thoughts, which often lead to action. For example, a problem marijuana user might believe that the drug provides the only way to relax. These users may interpret a situation as tense. The interpretation might activate the belief that they need marijuana to relax. This belief would likely lead to thoughts of using, which might inspire all the actions required to get high. In cognitive therapy, a client would learn to challenge beliefs in an effort to minimize drug use. Thus, the client may develop the skills to see situations as not so stressful and would certainly alter the belief that using drugs is the only great way to relax (Beck et al., 1993). Instead of smoking marijuana, one might listen to music, meditate, or exercise.

Therapists have developed many techniques for altering these beliefs. Most require identifying the underlying belief, and then looking for evidence to support or dispute it. One common strategy that cognitive-behavioral therapists employ includes Socratic questioning. Socratic questioning takes its name from Plato's descriptions of interactions between the ancient Greek philosopher Socrates and his students (Plato, 1999). Socrates rarely wasted time with didactic lectures. Instead, he guided students through a series of questions so that they might arrive at their own answers. Adept therapists might employ a comparable style of inquiry. Instead of merely providing information, this strategy teaches a process for discovery. Eventually, clients can learn to ask these sorts of questions of themselves so they can maintain sobriety without the therapist.

This process also elicits the thoughts and feelings most important to the client. For example, those who believe marijuana provides the only way to relax might respond particularly well to questions about alternative ways to unwind. Questions about restful recreation in general may prove helpful. Queries about favorite activities from before the clients began using marijuana might also work effectively. As clients generate their own list of preferred ways to soothe themselves without drugs, the belief that marijuana is the sole source of relaxation weakens. Note that clients would find their own examples more compelling than any list of calming circumstances the therapist might generate. This approach also respects the client's own ability to present evidence to alter beliefs (Overholser, 1987). Changing the thoughts about situations that previously led to drug use can help decrease problematic drug consumption.

Cognitive-behavioral therapy relies on other techniques too numerous to list here, but one key set of strategies concerns relapse prevention. Many people can quit using for a brief period but cannot maintain abstinence. Thus, many cognitive-behavioral techniques focus not only on quitting but also on avoiding the return to drugs. Thoughts and beliefs remain important in preventing relapse, given their relevance to a phenomenon known as the abstinence violation effect. The abstinence violation effect concerns the way people cope with backsliding once they have committed to altering their drug consumption.

Most people who decide to eliminate or decrease their use subsequently make mistakes. They use the drug when they intended to quit or use more than their established limit. The abstinence violation effect occurs when a small, thoughtless toke of a joint turns into a full weekend binge. It is as if people say "Well, I wrecked my abstinence, so I might as well use all that I can." Minimizing the impact of small slips is essential to relapse prevention. Although many believe that the pharmacology of the drug makes a single dose inevitably turn into a relapse, changes in thinking can actually prevent these slips from creating troubles. The interpretation of the slip appears to contribute more to relapse than the actual occurrence of the slip itself (Marlatt & Gordon, 1985).

No one doubts that intoxicated people can make poor decisions about continued drug use, and that the pharmacological effects of the drug certainly contribute to these decisions. Nevertheless, many who relapse report abstinence violation effects that occurred at extremely low doses of their drug of choice. Often a single sip of liquor or smell of marijuana led to decisions to binge. Pharmacology may not play a particularly strong role in these relapses. A crafty study by Marlatt, Demming, and Reid (1973) revealed that alcoholics who drank alcohol but did not know it did not show the abstinence violation effect. They did not continue drinking after the initial dose. In contrast, alcoholics given a placebo that they thought was alcohol did show the abstinence violation effect. They consumed considerably more alcohol after the placebo. Obviously, thoughts play an important role in relapse prevention.

A study like this one giving marijuana or a placebo to problem marijuana users has never been performed, in part because of ethical considerations. Yet at least one study of marijuana users in treatment supports the idea that interpretations of slips alter the chances of relapse. Seventy-five people from an abstinence-oriented treatment program participated. All had smoked marijuana after their designated quit dates. They completed questionnaires concerning their thoughts about these incidents of use. Three types of attributions appeared with the largest relapses: internal, stable, and global ones. Those people who thought that their mistaken use stemmed from their own internal lack of ability often continued using. If they thought lapses arose from consistent, stable qualities that applied in many settings, use also increased.

A few examples can help illustrate these different kinds of attribution. Suppose someone who had quit smoking marijuana attributed a lapse to a weak character. Weak character, if it exists at all, supposedly describes an internal quality about the person. Weak character is stable—it stays with the person in any situation. Thus, people who attribute a slip to an internal, stable, and global factor like weak character will likely relapse. In contrast, those who attributed their slips to external, unstable, specific circumstances faired better. If they thought the source of their lapses varied in an unstable way and did not apply to many settings, relapse proved less likely. For example, suppose someone attributed a lapse to an inability to refuse marijuana when it was offered. Refusing drugs is a skill that people can learn and that improves with practice. The absence of the skill probably relates to external circumstances—the lack of an opportunity to learn. The decreased skill remains unstable; it is possible to alter and improve it. In addition, people may not be offered drugs in every setting they encounter. This external, unstable, specific attribution is less likely to lead to relapse (Stephens, Curtin, Simpson, & Roffman, 1994).

Cognitive-behavioral therapy relies on the principles of learning theory to treat substance abuse problems. Clients can treat problem drug use as if it were identical to other learned behaviors. The treatment may work by altering beliefs about drug use and its consequences. It also focuses on the prevention of relapse by identifying situations that may increase the risk of drug use and then teaching alternative ways to act under those conditions. Relatively brief versions of the treatment have proven helpful for marijuana users in an empirical study, with about one-third of problem users eliminating drug-related problems. This treatment approach is not the only one with empirical support, however. Social support has proven equally successful in treating marijuana problems and may play a role in twelve-step programs.

Twelve-step Facilitation

This brief intervention helps individuals successfully join the twelve-step fellowship appropriate for their addictive behaviors. The fellowship consists of the recovering addicts who participate in the meetings and activities of the group. Unlike twelve-step programs, twelve-step facilitation is a form of psychotherapy. Twelve-step programs are not psychotherapy, since they employ no formal diagnoses or assessments. Instead, they use the 12 steps to solve problems, which include important processes like admitting powerlessness over the drug, developing a belief in something greater than oneself, taking a moral inventory of one's actions, making amends for one's wrongs, and carrying the message of recovery to other problem users.

The programs serve as opportunities to adopt a new, sober lifestyle that includes an active involvement in the fellowship. In this way, twelve-step approaches have many advantages over psychotherapy. Unlike most individual therapists, the fellowships provide free services in almost every major city in the world. They furnish a complete network of members, hot lines for telephone support, and the opportunity to serve the community. They also have a long tradition with many members reporting years of successful sobriety. These advantages inspired the development of twelve-step facilitation, which helps people tackle the initial tasks associated with joining the program (Nowinski & Baker, 1992).

Many mental health professionals recommend twelve-step meetings to clients with drug problems, but a simple suggestion is rarely enough to help someone connect to the fellowship. Twelve-step facilitation serves as a more formal way to help ensure a productive experience in the program. Data support twelve-step facilitation as an effective strategy for minimizing alcohol problems (Project MATCH, 1998). It would likely generalize to those experiencing negative consequences from marijuana.

In this treatment, each client meets regularly with a twelve-step facilitator who discusses the program, provides guidance about the steps, encourages active participation, and monitors progress. The facilitator is not a formal part of twelve-step programs. Twelve-step work does not require a coach, counselor, or therapist of any kind outside of the fellowship. But discussions with the facilitator may help decrease the chances of dropping out and may make the transition from new member to active member easier.

Membership in the fellowship enhances outcomes for alcoholics and likely helps problem users of cannabis (Morganstern, Labouvier, McCrady, Kahler, & Frey, 1997). Alcoholics Anonymous and twelve-step programs designed for drugs other than alcohol are quite similar. Although few formal studies address success rates, members of Narcotics Anonymous and Marijuana Anonymous report successful abstinence from marijuana. Mental health professionals recommend twelve-step treatment for people experiencing marijuana problems, suggesting considerable confidence in the program (Miller, Gold, & Pottash, 1989).

The success of social support treatment implies that the supportive aspects of Marijuana Anonymous or Narcotics Anonymous would help problem users. Perhaps generalizing from the studies of Alcoholics Anonymous is appropriate. All these programs rely on the 12 steps and the disease model. The central topics of the programs remain comparable, emphasizing spirituality, acceptance, and surrender. All the programs highlight the role of sponsors, peers, prayer, and regular attendance at meetings. Meetings of the groups follow similar formats. The corresponding processes within these groups suggest that twelve-step facilitation might work comparably regardless of which group a client chooses to join (Nowinski, 1996). Thus, twelve-step facilitation may show promise for problem marijuana users.

The Format of Facilitation

Just as no therapy works the same way twice, no two experiences in twelve-step programs are identical. Thus, twelve-step facilitation will function a little differently with each person. Nevertheless, the treatment focuses on common, core topics. These include an introduction to the program, acceptance, surrender, and getting active in the fellowship. Additional sessions can highlight elective topics. These include enabling—the role of others in facilitating drug consumption, and inventories—candid and thorough examinations of ethical transgressions. Conjoint sessions, where the therapist meets with both the problem user and meaningful people in his or her life, are also possible. These sessions often concentrate on enabling, too. The therapists need not be members of twelve-step programs, but they must have a familiarity with various meetings, an understanding of the disease model, and an appreciation for the role of spirituality in recovery. An extensive network of contacts with individuals who are active in the fellowship would also serve as an asset.

The initial session of twelve-step facilitation focuses on recommending a connection with the fellowship. If assessment reveals appropriate drug problems, the therapist requests regular attendance at treatment sessions and twelve-step meetings. In the initial stages of recovery, daily attendance at meetings serves as the goal. Programs traditionally suggest 90 meetings in the first 90 days of membership. The therapist also asks the client to keep a journal for monitoring attendance and reactions to these meetings. Readings of relevant publications related to the twelve-step group enhance treatment, too. Marijuana Anonymous publishes "Life with Hope" (1995), which details the steps and traditions of the program. Narcotics Anonymous publishes "Narcotics Anonymous" (1988), which describes comparable information about the steps and traditions. Subsequent sessions focus on reviews of the journal and readings, as well as assessments of any drug use or urges to use. The therapist offers praise for sober days, handling urges, and attendance at meetings. Treatment then turns to two central issues in twelve-step recovery: acceptance and surrender.

Acceptance concerns the tranquil understanding of personal limitations. It relates strongly to the first step of the program, admitting powerlessness over the drug. Some problem drug users report acceptance as a single, categorical shift in their thinking. Most experience a gradual change. The opposite of acceptance is denial—a resistance to the idea of powerlessness. Denial might include any thought that controlled drug use without problems remains an option. It serves as a natural reaction to any thought of personal limitation. Few people long to admit that controlled use of a substance remains impossible, particularly if they know others who claim to use drugs with impunity.

Frank discussions of the link between drug use and life's problems can help combat denial. Initially, users may misattribute problems that likely arose from drug use to other factors. As problem users grow more candid about how drugs have made their lives unmanageable, accepting powerlessness and personal limitations may grow easier. For example, people who smoke pot daily and experience a great deal of fatigue may attribute their exhaustion to trouble sleeping. After a few weeks in the program they may notice more energy, realize that daily smoking had drained their strength, and be more willing to accept that they cannot function optimally with the drug in their lives.

The acceptance of powerlessness over the drug leads to the difficult realization that one cannot function productively alone. This acceptance lays the groundwork for surrender. The realization of powerlessness essentially leads to an understanding that only something outside of one's self can provide appropriate support for sobriety. Essentially, acceptance of limitations leads to a request for help. Surrender includes a tranquil understanding of the importance of this outside support and sets the stage for the idea of a higher power. This idea remains one of the more novel and controversial aspects of twelve-step work.

The higher power is unique to these twelve-step programs; no other approach to substance abuse treatment relies on it. It creates considerable resistance in some, despite the twelve-step program's pluralistic approach. References to the higher power always emphasize the individual's personal understanding. There is no recommended higher power. Some members view the fellowship as their higher power; some rely on the deities they grew up learning about in organized religion. Others view love or knowledge as something separate from themselves that can assist them (Wallace, 1996). The signs of surrender generally include an improved ability to accept outside help. This ability often manifests in increased involvement with the fellowship.

Later sessions in twelve-step facilitation focus on increasing involvement with the fellowship. Getting active requires consistent attempts to understand the program. Regular attendance and participation at meetings serve as a first step toward increased activity. Attendance and participation at different types of meetings serve as signs of commitment to a sober lifestyle. In the initial stages of recovery, people often report mixed feelings about their attendance. No two twelve-step meetings are identical. Clients who dislike meetings in one location can attend those in another. Some meetings focus on the steps, the experience of particular speakers, or the open comments of any member willing to talk. Meetings designed for homogeneous groups also exist. For example, interested clients may find gatherings that include only men, women, Latinos, or members of the gay community.

The facilitator and client can discuss reactions to the meetings, which often reveal a great deal about attitudes and beliefs related to drugs and treatment. Impressions of meetings often improve with increased participation. Simple attendance serves as a great start, but speaking up at meetings is essential to increased participation. Clients can discuss their concerns about speaking with the facilitator and even role play making comments.

Connecting to a sponsor and twelve-step peers also enhances involvement. A sponsor, someone with more experience who is willing to help take the message of the program to new members, can answer questions and provide advice about building a meaningful life without drugs. Other members of the program can serve as peers in the recovery process, perhaps providing a supportive, interested phone call or a ride to a meeting. These connections help form a new social support network committed to a different lifestyle. Clients initially may feel reluctant to ask for phone numbers, much less sponsorship. These actions may require a genuine acknowledgment of the need for outside connections. The therapist may suggest that the client talk with many members after meetings, gradually growing more familiar with more people, and then requesting phone numbers and sponsorship as comfort increases. Role plays may help in these situations, too.

Reading about the program also helps increase understanding and involvement. Numerous books and pamphlets in the twelve-step literature address complex topics like acceptance, surrender, and powerlessness. The readings also reveal the stories of many individuals who experienced negative consequences from drugs and turned to the fellowship for help. These stories often move reader's emotions, enhancing a sense of connection to other problem users. Candid discussions of these readings with the facilitator and members of the fellowship can increase this sense of connection. This process can help problem users make the transition from an intellectual comprehension of key concepts of the program to a more thorough, experiential understanding. The combination of all these ideas and techniques can lead to a lifelong commitment to a new way of living, providing social support, spirituality, and a strategy for ending the negative consequences of drug use (Nowinski & Baker, 1992).

Motivational Interviewing

Motivational interviewing relies on brief interactions with a therapist to help clients decrease problems. The treatment enhances motivation before attempting any changes in behavior. Therapists adopt this approach because in the absence of a client's motivation, any efforts to teach techniques for limiting drug use simply waste time. Once a client's motivation has increased, strategies for eliminating drug problems have a better chance of success. Motivational interviewing focuses on identifying the client's own reasons to quit. Once these reasons help increase desire, clients often develop their own strategies for eliminating problem drug use. Many people stop using drugs on their own; motivational interviewing essentially enhances the chances that a client will join this group. The social interaction with the therapist may highlight the negative consequences of drug use, leading people to change their own lives.

Motivational interviewing relies on a couple of general principles to help clients decrease marijuana problems. First, the therapist behaves in ways that increase the likelihood of change, such as listening attentively without judgment or blame. Second, the therapist employs the stages of change model. This model provides a view of change as a process that requires different interventions for different stages of the client's willingness to act. The general behaviors most likely to induce change involve empathy, nonpossessive warmth, and genuineness. Carl Rogers originally emphasized these attributes in the treatment he invented, client-centered therapy (Rogers, 1950). These therapist actions appear commonly in successful treatments. The presence of these behaviors in many treatments may serve as a good explanation for why different therapies produce comparable results (Wampold et al., 1997).

Although everyone has an implicit feel for empathy, nonpossessive warmth, and genuineness, these qualities prove difficult to define in the abstract. Empathy concerns the ability to identify with another person's feelings. Empathic reactions clearly indicate that the therapist understands the client's view of situations. This understanding and empathy have great importance in the treatment of substance abuse. Therapists may have never experienced each client's situation exactly, but they are certainly familiar with frustration, disappointment, sadness, and the range of emotions that accompany change. Expressions of this empathy enhance the relationship between client and therapist. This sharing of feeling may increase the client's trust, encouraging candid disclosures. Sorting through the ambivalent and conflicted feelings associated with drug problems may help clients make clear decisions about decreasing the negative consequences of their use.

Nonpossessive warmth refers to a therapist's interactive style. Warmth suggests a generally good-natured approach to therapy and a sincere appreciation of the client's intricacies and uniqueness. The nonpossessive aspect implies that the therapist does not withdraw, cajole, or manipulate. The warmth does not disappear and reappear with changes in behavior. Thus, clients need not fear a bad reaction if they report emotions or behaviors they consider negative. The therapist's style should not change if the client grows upset, reports urges, or uses drugs. Demonstrations of warmth vary among therapists as they do among other people. Nevertheless, a sincere smile, an attentive nod, and considerate listening invariably enhance interactions and reveal warmth.

Genuineness appears in authentic, trustworthy, realistic behavior. Clients rely on sincere reactions that are free from pretense or affectation. A therapist who seems natural creates a more comfortable atmosphere than one who appears scripted, stilted, or phony. Therapists who show genuineness have body language, eye contact, and facial expressions that correspond to their words. Essentially, the human interaction should feel more important than following a treatment protocol, as this quality enhances rapport. Clients of therapists who show genuine interactions report feeling that they are getting to know the therapist, in the sense that they relate to each other rather than simply exchange information. Although no one can point to specific actions out of context and label them genuine or not, people easily identify therapists who seem consistent, true to themselves, and real (Miller & Rollnick, 1991).

The Stages of Change Model

This empathy, warmth, and genuineness lay the foundation for any productive, therapeutic interaction. Many therapies rely on these aspects of the relationship to help support growth. Motivational interviewing combines these qualities with the stages of change model to help inspire a shift away from problem drug use. The stages of change model describes particular steps that individuals appear to take any time they alter a problem behavior (Prochaska & DiClemente, 1983). Researchers identified these steps by interviewing people who quit smoking on their own. The investigators proposed six stages common to the process: precontemplation, contemplation, determination, action, maintenance, and relapse (Prochaska, Norcross, & DiClemente, 1994).

Precontemplation describes the period before individuals consider altering their behavior. The idea of precontemplation as a stage of change may serve as one of the most novel aspects of this model. Marijuana users in precontemplation have never considered changing their consumption. An adept therapist would not waste time attempting to teach these people how to quit using; they would likely lack motivation to learn these skills. Instead, the therapist would begin with assessment. A report on the amount and frequency of marijuana use would serve as a good start. The therapist would also want to ask about any associated consequences, including negative emotions, fatigue, uncomfortable interpersonal interactions, or any other negative consequences the smoker might experience. This assessment can often make the connections between use and consequences more salient. If these connections lead individuals to consider change in any way, they have entered the contemplation stage.

Contemplation includes the weighing of the pros and cons of altering actions or continuing the same behavior. The motivational interviewer would allow the marijuana smokers to candidly report all the positive experiences they attributed to drug use, including any beliefs about enhanced sexual interactions, enjoyment, slowing of time, or connections to the counterculture. Then the interviewer might ask smokers to highlight negative consequences. Initial assessments of pros and cons often reveal strong desires to continue using, as well as equally strong desires to stop. This situation may reflect the ambivalence people feel about altering their consumption of marijuana. Ambivalence serves as a common and important component of contemplation. Other approaches to treatment may see this ambivalence as denial. The stages of change model emphasizes ambivalence as an inherent part of change. During further discussion, the therapist respectfully reflects the marijuana users's concerns back to them, emphasizing the negative consequences that they generated earlier. This process often leads problem users to a decision to change. A firm decision to change qualifies as a step toward determination.

Determination begins with a clearly stated desire to alter actions. This stage serves as the appropriate time for a marijuana user to formulate a plan for limiting consumption. Note that any attempts to devise a strategy for change before the determination stage would essentially waste effort. Motivation must increase before a plan can succeed. The plan often stems from brainstorming between the interviewer and the smoker and may include any options that look promising. For example, the strategy for change may rely on techniques from cognitive-behavioral therapy like altering beliefs and preventing relapse. In addition, the smoker may decide that membership in a twelve-step program sounds appropriate.

Once clients regularly alter old behaviors in favor of new ones, they have entered the action stage. They no longer merely consider change; they actually do it. This stage proves particularly informative. The genuine experience of new habits and actions can reveal valuable information unanticipated during contemplation and determination. Clients may find some situations easier than they expected. Other aspects of abstinence or controlled use may prove unexpectedly difficult. The motivational interviewers will now offer reassurance about the process becoming less difficult with practice. They will help clients solve problems related to use. They will listen attentively to detailed descriptions of difficulties and proud retellings of each resisted temptation.

After a steady period of action, clients may report increased confidence in their skills. This sense of efficacy, an optimism in one's own ability to continue the new behaviors, serves as a hallmark of the maintenance stage. Self-efficacy and sustained change are the keys to maintenance. Client and therapist will work together now to prevent relapse. They will identify situations that put the smoker at high risk for relapse and plan ways to avoid problematic use in these circumstances. For example, clients may decide to avoid parties where drugs are present. They may role play refusing drugs if they are offered. They may practice relaxation techniques if tension often preceded their drug use. They may call a hot line or a friend in times of temptation. Note that these techniques for preventing relapse are consistent with twelve-step and cognitive-behavioral approaches. Perhaps this overlap contributes to the comparable results of these different programs.

Occasional backsliding occurs in many efforts to alter behavior. Original studies of people who quit smoking cigarettes reveal that they rarely remain abstinent on their first try (Prochaska et al., 1994). They quit, relapse, and quit again. The stages of change model considers lapses and relapses as another category of change. This approach may help normalize the occasional slip. Considering lapses as a part of the change process may decrease the chances of an abstinence violation effect, transforming a slip into a full-blown relapse. The key to the lapse stage parallels the key to the maintenance stage: preventing relapse. Lapses require immediate action. Lapsing smokers can prevent relapse by rapidly exiting the situation and removing the chance of continued use.

Many who lapse berate themselves, but their time and energy may be better spent identifying the precursors to the drug use. A frank examination may reveal a new high-risk situation, providing the opportunity to formulate a plan for how to handle this predicament in the future.

For example, a former cannabis smoker may find himself lighting up after a fight with a family member. This situation may not be one that he had identified as high risk before. Now he knows that he needs to plan new ways to deal with conflict. He can turn this lapse into a learning experience to prevent later use. Thus, lapses remain a part of the change process; planning for them may minimize problems. By combining good therapeutic skills in general and targeted interventions for each stage of change, a motivational interviewer can help problem drug users through many steps toward minimizing problem drug use.

Conclusions

People who experience negative consequences from marijuana have a number of imperfect but promising alternatives for eliminating problems. Some may alter their consumption of the drug on their own. Those who choose to enter treatment have a few options. One study of twelve-session group treatments reveals that approximately one-third of the participants could eliminate marijuana problems, and 14% remained abstinent from the drug for a year. Data from studies of other drugs of abuse suggest that longer treatments may improve the outcomes of therapy. Pharmacological and psychological treatments that address any coexisting disorders, such as anxiety or depression, might also help limit marijuana-related harm. Three promising substance abuse therapies may also work well with marijuana problems: cognitive-behavioral therapy, twelve-step facilitation, and motivational interviewing. Although no treatment is perfect, with considerable effort and hard work, motivated people can eliminate their use of this drug and minimize its negative effects.

12
Final Thoughts

A great deal of the available information on marijuana appears in this book. But the marijuana literature is extensive. It does not lend itself to easy summaries or interpretations. As mentioned in the preface, any attempt to explain this research may say more about the explainer than the explained. People who claim to be rational often gather information before forming opinions and making decisions. Others form opinions, make decisions, and then go in search of reasons afterward. A selective reading of this research can buttress nearly any argument for or against the drug. A careful reading, however, reveals several consistent themes.

A few points about marijuana remain unarguable. The plant is at least 10,000 years old. Its medicinal applications began at least 4,500 years ago. Recreational use has also been around for thousands of years. Cannabis is the most popular illicit drug in the world. Hundreds of millions of people have tried it. Only a small fraction of them develop problems with other illicit drugs. Less than one-tenth of the people who ever try marijuana end up using it regularly. Fewer still develop troubles with it. Some fix the problems on their own. Many respond well to therapy. Current treatments are promising, but not perfect.

A few facts about marijuana intoxication also seem clear. The experience is difficult to depict and varies dramatically from person to person and across situations. Some people feel more relaxed, happy, and alive. Others feel paranoid and anxious. After smoking marijuana, people experience time, space, and emotions differently. They eat more and crave sweets. Intoxicated people do not learn new material well. They cannot solve complex problems quickly, and their brain waves change. They can drive a car as well as the unintoxicated, but these consistent results are so counterintuitive that most people find them unbelievable. Individuals are no more aggressive after smoking marijuana. Intoxication usually last a couple of hours, depending upon dosage. After it ends, there is little hangover or residual effect.

Several points about chronic use are also evident. After years of daily smoking, people do not show any changes in brain structure, unless they started using the drug before adulthood. They also rarely show deficits on standard measures of intelligence, thinking, or ability. Yet sensitive tests show changes in brain function. Chronic users can perform more poorly on complex, difficult tasks that require fast reactions and focused attention. The practical implications of these findings continue to generate debate. Studies of chronic users have yet to reveal dramatic health problems, but their lungs show changes that suggest an increased risk for cancer. Chronic users do not show a consistent, identifiable, amotivational syndrome. Yet people who are high all the time probably do not get a great deal of work done. Compared to alcohol, cigarettes, and over-the-counter medications, occasional marijuana use causes little harm.

The future for cannabis holds many possibilities. Research on the cannabinoids and their receptors will undoubtedly continue to tell more about the human mind and body. This work could reveal additional information about the brain and immune system. Further work can test the efficacy of marijuana and the cannabinoids as medical treatments. Data on the long-term health effects of the drug could address many unanswered questions, particularly those concerning the lung and brain. Techniques may evolve to limit the drug's negative consequences, like recent efforts to develop the vaporizer to reduce noxious components of smoke. Treatments for problem users could improve. In addition, laws related to the drug may change.

Cannabis became essentially illegal in the United States in 1937. Despite over 60 years of prohibition, more people have used the drug than ever before. Police arrest over half a million Americans each year for crimes related to marijuana. Government spends billions annually on marijuana control. Several authors suggest that alternative policies may prove cheaper, send fewer people to jail, and maintain respect for the law. After reviewing this literature, readers may agree. These plans include legalization, decriminalization, and licensing users. The experiences of other countries suggest little change in levels of drug problems after decriminalization, but many people feel that these experiences would not apply in the United States. Nearly a dozen states have decriminalized the drug at one time or another with little impact on use. Federal prohibition remained, however, which may have kept rates of consumption down. Yet most people who do not use the drug claim they have no interest regardless of its legal status. Changes in legislation have the potential to save money and increase respect for the law. They also might have unforeseeable long-term consequences.

Cannabis policy reflects ideas about the drug. Many claim that the laws have developed in an effort to limit harm. The extent of this harm is an empirical question. Studies can establish the drug's negative impact. Relevant research appears throughout this book. The harm does not appear dramatic and may not justify current policy. Other contributors to marijuana laws may not be so easy to address with research. Marijuana problems are not the only reason for these policies. The laws may reveal unspoken attitudes about the people who want to alter consciousness. Some citizens may view people who want to change their consciousness as evil or bad. These views may become the strongest determinants of cannabis policy, particularly as news of the drug's limited harm reaches everyone.

Each of us carries tacit assumptions about thoughts, awareness, and an internal life. It is perfectly human to assume that consciousness should be a certain way. We all have stereotypes about techniques designed to alter thoughts and awareness. Mental activities for changing consciousness can provide examples. Look at common attitudes about meditation, frequent prayer, and hypnosis. People who do not engage in these practices may see those who do as decidedly deviant. Yet the activities remain legal, perhaps because they cause no harm. Physical activities also change consciousness. Skydiving, bungee jumping, and motorcycling remain legal. In addition, they are not harmless. But more people have died from these activities than ever overdosed on marijuana. Many see these actions as deviant. Yet they do not bother people enough to inspire prohibition. Something about the required effort and the attempts at safety seem to keep these activities from becoming crimes. The fact that they do not require external chemicals may help.

A few chemicals that alter consciousness also remain accepted. Caffeine, nicotine, and alcohol have established effects on thoughts and moods. Their toxicity is much higher than marijuana's. Yet as long as they are consumed in a way that causes no harm to others, they remain legal. Caffeine and nicotine do not produce the dramatic changes in consciousness that seem to create concern. Alcohol alters thoughts more dramatically, which may explain previous American attempts to prohibit it. Yet alcohol's familiarity, popularity, and potential to generate tax revenues during an economic depression caused prohibition's repeal. Alcohol consumption still has its detractors, and the drug creates plenty of harm. Alcoholics certainly suffer from associated stigma. With appropriate economic incentive, however, voters seemed willing to accept this chemical way to alter consciousness. Those who use alcohol responsibly seem to suffer little harm from the drug or from the legislation that controls it. Citizens seem to trust adults enough to let them attempt to use this drug in a way that will not cause problems. Can we extend this trust to people who use marijuana?

From a combination of economic incentives and a sense of justice, the world has slouched toward progress in appreciating diversity. People are starting to respect each other a little more, regardless of age, ethnicity, gender, occupation, sexual orientation, religion, political affiliation, or education. Many argue that this greater respect benefits everyone. We approach a point where people might tolerate others who think differently. Perhaps we could tolerate people who want to use marijuana without causing harm to themselves or others.

References

A. C. Nielson Co. (1998). TV statistics. Available: http://www.tvta.org/stats.html.

Abel, E. (1977). The relationship between cannabis and violence: A review. *Psychological Bulletin, 84,* 193–211.

Abel, E. (1980). *Marijuana: The first twelve thousand years.* New York: Plenum.

Abrahamov, A., Abrahamov, A., & Mechoulam, R. (1995). An efficient new cannabinoid antiemetic in pediatric oncology. *Life Sciences, 56,* 2097–2102.

Adamec, C., Pihl, R. O., & Leiter, L. (1976). An analysis of the subjective marijuana experience. *International Journal of the Addictions, 11,* 295–307.

Adams, A. J., Brown, B., Haegerstrom-Portnoy, G., & Flom, M. C. (1976). Evidence for acute efffects of alcohol and marijuana on color discrimination. *Perception and Psychophysics, 20,* 119–124.

Adelman, S. A., & Weiss, R. D. (1989). What is therapeutic about inpatient alcoholism treatment? *Hospital and Community Psychiatry, 40,* 515–519.

Alcott, L. M. (1869/1976). *Plots and counterplots: More unknown thrillers of Louisa May Alcott.* M. Stern (Ed.). New York: William Morrow.

Aldrich, M. R. (1997). History of therapeutic cannabis. In M. L. Mathre (Ed.), *Cannabis in medical practice* (pp. 35–55). London: McFarland.

Aldrich, M. R., & Mikuriya, T. (1988). Savings in California marijuana law enforcement costs attributable to the Moscone Act of 1976—A summary. *Journal of Psychoactive Drugs, 20,* 75–81.

Ali, R., Christie, P., Hawks, D., Lenton, S., Hall, W., Donnelly, N., et al. (1998). *The social impacts of the cannabis expiation notice scheme in South Australia.* Canberra, Australia: Department of Health and Family Services.

American Management Association. (1998). *Drug testing and monitoring survey*. New York: Author.

American Psychiatric Association (APA). (1952). *Diagnostic and statistical manual of mental disorders* (1st ed.). Washington, DC: Author.

American Psychiatric Association (APA). (1968). *Diagnostic and statistical manual of mental disorders* (2nd ed.). Washington, DC: Author.

American Psychiatric Association (APA). (1994). *Diagnostic and statistical manual of mental disorders* (4th ed.). Washington, DC: Author.

Ames, F. R. (1986). Anticonvulsant effect of cannabidiol. *South African Medical Journal, 69,* 14.

Andreasson, S., Allebeck, P., & Rydberg, U. (1989). Schizophrenia in users and non-users of cannabis. *Acta Psychiatrica Scandinavica, 79,* 505–510.

Andreoli, T. E., Carpenter, C. C., Bennet, C. J., & Plum, F. (Eds.). (1997). *Cecil essentials of medicine*. Philadelphia: Saunders.

Armor, D. J., Polich, J. M. & Stambul, H. B. (1978). *Alcoholism and treatment*. New York: Wiley.

Baker, D., Pryxe, G., Croxford, J. L., Brown, P., Pertwee, R. G., Huffman, J. W., et al. (2000). Cannabinoids control spasticity and tremor in a multiple sclerosis model. *Nature, 404,* 84–87.

Balzac, H. (1900). *Letters to Madame Hanska*. Boston: Little, Brown.

Basavarajappa, B. S., & Hungund, B. L. (1999). Chronic ethanol increases the cannabinoid receptor agonist anandamide and its precursor N-arachidonoylphosphatidylethanolamine in SK-N-SH cells. *Journal of Neurochemistry, 72,* 522–528.

Basu, D., Malhotra, A., Bhagat, A., & Varma, V. K. (1999). Cannabis psychosis and acute schizophrenia: A case control study from India. *European Addiction Research, 5,* 71–73.

Bates, M. N., & Blakely, T. A. (1999). Role of cannabis in motor vehicle crashes. *Epidemiological Reviews, 21,* 222–232.

Baudelaire, C. (1861/1989). *The flowers of evil.* M. Mathews & J. Mathews (Eds.). New York: New Directions.

Beal, J. E., Olson, R., Laubenstein, L., Morales, J. O., Bellman, P., Yangco, B., et al. (1995). Dronabinol as a treatment for anorexia associated with weight loss in patients with AIDS. *Journal of Pain and Symptom Management, 10,* 89–97.

Beal, J. E., Olson, R., Lefkowitz, L., Laubenstein, L., Bellman, P., Yangco, B., et al. (1997). Long-term efficacy and safety of dronabinol for Acquired Immunodeficiency Syndrome–associated anorexia. *Journal of Pain and Symptom Management, 14,* 7–14.

Bech, P., Rafaelsen, L., & Rafaelsen, O. J. (1973). Cannabis and alcohol: Effects on estimation of time and distance. *Psychopharmacologia, 32,* 373–381.

Beck, A. T., Wright, F. D., Newman, C. F., & Liese, B. S. (1993). *Cognitive therapy of substance abuse.* New York: Guilford.

Bell, J. (1857). On the haschish or Cannabis Indica. *Boston Medical and Surgical Journal, 56,* 209–216.

Bello, J. (1996). *The benefits of marijuana: Physical, psychological, and spiritual.* Boca Raton, FL: Lifeservices.

Benet, S. (1975). Early diffusion and folk uses of hemp. In V. Rubin (Ed.), *Cannabis and culture* (pp. 39–50). The Hague: Mouton.

Benjamin, D. K., & Miller, R. L. (1991). *Undoing drugs.* New York: Basic Books.

Bennett, W. (1991). The plea to legalize drugs is a siren call to surrender. In M. Lyman & G. Potter (Eds.), *Drugs in society* (p. 339). Cincinnati: Anderson.

Bidaut-Russell, M., Devane, W. A., & Howlett, A. C. (1990). Cannabinoid receptors and modulation of cyclic AMP accumulation in the rat brain. *Journal of Neurochemistry, 55,* 21–55.

Bishop, J. L. (1966/1868). *A history of American manufactures.* New York: Kelley.

Blank, D., & Fenton, J. (1991). Early employment testing for marijuana. In S. Guse & J. Walsh (Eds.), *Drugs in the workplace* (monograph 91, pp. 151–167). Rockville, MD: National Institute on Drug Abuse.

Blaze-Temple, D., & Lo, S. K. (1992). Stages of drug use: A community survey of Perth teenagers. *British Journal of Addiction, 87,* 215–225.

Bleiberg, J. L., Devlin, P., Croan, J., & Briscoe, R. (1994). Relationship between treatment length and outcome in a therapeutic community. *International Journal of the Addictions, 29,* 729–740.

Block, R. I., Erwin, W. J., Farinpour, R., & Braverman, K. (1998). Sedative, stimulant, and other subjective effects of marijuana: Relationships to smoking techniques. *Pharmacology, Biochemistry and Behavior, 59,* 405–412.

Block, R. I., Farinpour, R., & Braverman, K. (1992). Acute effects of marijuana on cognition: Relationships to chronic effects and smoking techniques. *Pharmacology, Biochemistry and Behavior, 43,* 907–917.

Block R. I., & Ghoneim, M. M. (1993). Effects of chronic marijuana use on human cognition. *Psychopharmacology, 110,* 219–228.

Block, R. I., O'Leary, D. S., Ehrhardt, J. C., Augustinack, J. C., Ghoneim, M. M., Arndt, S., et al. (2000). Effects of frequent marijuana use on brain tissue volume and composition. *NeuroReport, 11,* 491–496.

Block, R. I., O'Leary, D. S., Hichwa, R. D., Augustinack, J. C., Ponto, L. L. B., Ghoneim, M. M., et al. (2000). Cerebellar hypoactivity in frequent marijuana users. *NeuroReport, 11*, 749–753.

Block, R. I. & Wittenborn, J. R. (1984a). Marijuana effects on semantic memory: Verification of common and uncommon category members. *Psychological Reports, 55*, 503–512.

Block, R. I. & Wittenborn, J. R. (1984b). Marijuana effects on visual imagery in a paired-associate task. *Perceptual & motor skills, 58*, 759–766.

Block, R. I., & Wittenborn, J. R. (1986). Marijuana effects on the speed of memory retrieval in the letter-matching task. *International Journal of the Addictions, 21*, 281–285.

Bloom, J. W., Kaltenborn, W. T., Paoletti, P., Camilli, A., & Leibowitz, M. S. (1987). Respiratory effects of non-tobacco cigarettes. *British Medical Journal, 295*, 516–518.

Blum, R. H. (1984). *Handbook of abusable drugs*. New York: Gardner.

Boire, R. G. (1992). *Marijuana law*. Berkeley, CA: Ronin.

Bonnie, R. J., & Whitebread, C. H. (1974). The marijuana conviction: A history of marihuana prohibition in the United States. Charlottesville: University Press of Virginia.

Borg, J., Gershon, S., & Alpert, M. (1975). Dose effects of smoked marijuana on human cognitive and motor functions. *Psychopharmacologia, 42*, 211–218.

Bornheim, L. M., Kim, K. Y., Li, J., Perotti, B. Y., & Benet, L. Z. (1995). Effect of cannabidiol pretreatment on the kinetics of tetrahydrocannabinol metabolites in mouse brain. *Drug Metabolism & Disposition, 23*, 825–831.

Bowman, M., & Pihl, R. O. (1973). Cannabis: Psychological effects of chronic heavy use: A controlled study of intellectual functioning in chronic users of high potency cannabis. *Psychopharmacologia, 29*, 159–170.

Braden, W., Stillman, R. C., & Wyatt, R. J. (1974). Effects of marijuana on contingent negative variation and reaction time. *Archives of General Psychiatry, 31*, 537–541.

Brazis, M. Z., & Mathre, M. L. (1997). Dosage and administration of cannabis. In M. L. Mathre (Ed.), *Cannabis in medical practice* (pp. 142–156). London: McFarland.

Brecher, E. M. (1972). *Licit and illicit drugs*. Boston: Little, Brown.

British Medical Association. (1997). Therapeutic uses of cannabis. Amsterdam: Harwood.

Bromberg, W. (1939). Marihuana: A psychiatric study. *Journal of the American Medical Association, 113*, 4–12.

Bromberg, W., & Rodgers, T. C. (1946). Marihuana and aggressive crime. *American Journal of Psychiatry, 102,* 825–827.

Brown, E. J., Flanagan, T. J., & McLeod, M. (Eds.). (1984). *Sourcebook of criminal justice statistics–1983.* Washington DC: U.S. Government Printing Office, U.S. Department of Justice, Bureau of Justice Statistics.

Brown, S. A. (1993). Recovery patterns in adolescent substance abuse. In J. S. Baer, G. A. Marlatt, & R. J. McMahon (Eds.), *Addictive behaviors across the life span* (pp. 161–183). Newbury Park: Sage.

Bruce, T. J., Spiegel, D. A., & Hegel, M. T. (1999). Cognitive-behavioral therapy helps prevent relapse and recurrence of panic disorder following alprazolam discontinuation: A long-term follow-up of the Peoria and Dartmouth studies. *Journal of Consulting and Clinical Psychology, 67,* 151–156.

Bruera, E., & Higginson, I. (Eds.). (1996). *Cachexia-anorexia in cancer patients.* New York: Oxford University Press.

Bureau of Census. (1999). *Statistical abstract of the U.S.* Washington, DC: Congressional Information Services.

Burish, T. G., & Tope, D. M. (1992). Psychological techniques for controlling the adverse side effects of cancer chemotherapy: Findings from a decade of research. *Journal of Pain and Symptom Management, 7,* 287–301.

Burke, J. (1999). It's not how hard you work but how you work hard: Evaluating workaholism components. *International Journal of Stress Management, 6,* 225–239.

Burton, R. (1621/1977). *Anatomy of melancholy.* New York: Vintage.

Cabral, G. A. (1999). Cannabinoid receptors in sperm. In G. G. Nahas, K. M. Sutin, D. J. Harvey, & S. Agurell (Eds.), *Marijuana and medicine* (pp. 317–326). Totowa, NJ: Humana.

Callahan, E. J. (1980). Alternative strategies in the treatment of narcotic addiction: A review. In W. R. Miller (Ed.), *The addictive behaviors* (pp. 143–168). New York: Pergamon.

Cami, J., Guerra, D., Ugena, B., Segura, J., & De La Torre, R. (1991). Effects of subject expectancy on THC intoxication and disposition from smoked hashish cigarettes. *Pharmacology, Biochemistry and Behavior, 40,* 115–119.

Campbell, A. M. G., Evans, M., Thomson, J. L. G., & Williams, M. J. (1971). Cerebral atrophy in young cannabis smokers. *Lancet, 2,* 1219–1224.

Canadian Centre on Substance Abuse. (1998). *Cannabis control in Canada: Options regarding possession.* Ottawa, Canada: Author.

Caplan, P. J. (1995). *They say you're crazy.* Reading, MA: Addison-Wesley.

Cappell, H. D., & Pliner, P. L. (1973). Volitional control of marijuana intoxication: A study of the ability to "come down" on command. *Journal of Abnormal Psychology, 82,* 428–434.

Carlin, A. S., Bakker, C. B., Halpern, L., & Post, R. D. (1972). Social facilitation of marijuana intoxication: Impact of social set and pharmacological activity. *Journal of Abnormal Psychology, 80,* 132–140.

Carlin, A. S., Post, R. D., Bakker, C. B., & Halpern, L. M. (1974). The role of modeling and previous experience in the facilitation of marijuana intoxication. *Journal of Nervous and Mental Disease, 159,* 275–281.

Carlin, A. S., & Trupin, E. W. (1977). The effect of long-term chronic marijuana use on neuropsychological functioning. *International Journal of the Addictions, 12,* 617–624.

Carlini, E. A., & Cunha, J. M. (1981). Hypnotic and antiepileptic effects of cannabidiol. *Journal of Clinical Pharmacology, 21,* 417S–427S.

Carrier, L. (1962). *The beginnings of agriculture in America.* New York: Johnson Reprint Co.

Carter, T. F. (1968). *The invention of paper in China.* New Haven: Yale University Press.

Carter, W. E. (1980). *Cannabis in Costa Rica.* Philadelphia: Institute of the Study of Human Issues.

Caspari, D. (1999). Cannabis and schizophrenia: Results of a follow-up study. *European Archives of Psychiatry and Clinical Neuroscience, 249,* 45–49.

Casswell, S. (1975). Cannabis intoxication: Effects of monetary incentive on performance, a controlled investigation of behavioural tolerance in moderate users of cannabis. *Perceptual and Motor Skills, 41,* 423–434.

Casswell, S., & Marks, D. F. (1973). Cannabis and temporal disintegration in experienced and naive subjects. *Science, 179,* 803–805.

Casto, D. M. (1970). Marijuana and the assassins—An etymological investigation. *International Journal of the Addictions, 5,* 747–757.

Center on Addiction and Substance Abuse (CASA). (1996). National survey of American attitudes on substance abuse: II. Teens and their parents. New York: CASA at Columbia University.

Chait, L. D. (1990). Subjective and behavioral effects of marijuana the morning after smoking. *Psychopharmacology, 100,* 328–333.

Chait, L. D., Fischman, M. W., & Schuster, C. R. (1985). Hangover effects the morning after marijuana smoking. *Drug & Alcohol Dependence, 15,* 229–238.

Chait, L. D., & Pierri, J. (1992). Effects of smoked marijuana on human performance: A critical review. In L. Murphy & A. Bartke (Eds.), *Marijuana/cannabinoids: Neurobiology and neurophysiology* (pp. 387–423). Boca Raton: CRC.

Chandler, L. S. Richardson, G. A., Gallagher, J. D., & Day, N. L. (1996). Prenatal exposure to alcohol and marijuana: Effects on motor development of preschool children. *Alcoholism: Clinical and Experimental Research, 20,* 455–461.

Chang, K. (1968). The archeology of ancient China. New Haven: Yale University Press.

Chen, J., Marmur, R., Pulles, A., Paredes, W., & Gardner, E. L. (1993). Ventral tegmental microinjection of delta-9-tetrahydrocannabinol enhances ventral tegmental somatodendritic dopamine levels but not forebrain dopamine levels: Evidence for local neural action by marijuana's psychoactive ingredient. *Brain Research, 621,* 65–70.

Cherek, D. R., Moeller, F. G., Schnapp, W., & Dougherty, D. M. (1997). Studies of violent and nonviolent male parolees: I. Laboratory and psychometric measurements of aggression. *Biological Psychiatry, 41,* 514–522.

Cherek, D. R., Roache, J. D., Egli, M. Davis, C., et al. (1993). Acute effects of marijuana smoking on aggressive escape, and post-maintained responding of male drug users. *Psychopharmacology, 111,* 163–168.

Chopra, I. C., & Chopra, R. N. (1957). The use of cannabis drugs in India. *Bulletin on Narcotics, 1,* 4–29.

Chowdhury, A. N., & Bagchi, D. J. (1993). Koro in heroin withdrawal. *Journal of Psychoactive Drugs, 25,* 257–258.

Chowdhury, A. N., & Bera, N. K. (1994). Koro following cannabis smoking: Two case reports. *Addiction, 89,* 1017–1020.

Clark, L. D., & Nakashima, E. N. (1968). Experimental studies of marijuana. *American Journal of Psychiatry, 125,* 379–384.

Clark, W. C., Janal, M. N., Zeidenberb, P., & Nahas, G. (1981). Effects of moderate and high doses of marijuana on thermal pain: A sensory decision analysis. *Journal of Clinical Pharmacology, 21,* 299S–310S.

Clarke, R. C. (1998). *Hashish!* Los Angeles: Red Eye.

Clifford, D. B. (1983). Tetrahydrocannabinol for tremor in multiple sclerosis. *Annals of Neurology, 13,* 669–671.

Co, B. T., Goodwin, D. W., Gado, M., Mikhael, M., & Hill, S. Y. (1977). Absence of cerebral atrophy in chronic cannabis users: Evaluation by computerized transaxial tomography. *Journal of the Amereican Medical Association, 237,* 1229–1230.

Coates, R. A., Farewell, V. T., Raboud, J., Read, S. E., MacFadden, D. K., Calzavara, L. M., et al. (1990). Cofactors of progression to acquired immunodeficiency syndrome in a cohort of male sexual contacts of men with immunodeficiency virus disease. *American Journal of Epidemiology, 132,* 717–722.

Cohen, J. (1990). *Statistical power analysis for the behavioral sciences*. Hillsdale, NJ: Lawrence Erlbaum.

Cohen, M. J., & Rickles, W. H., Jr. (1974). Performance on a verbal learning task by subjects of heavy past marijuana usage. *Psychopharmacologia, 37*, 323–330.

Cohen, S. (1976). The 94 day cannabis study. *Annals of the New York Academy of Sciences, 282*, 211–220.

Cohen, S. (1979). Marihuana: A new ball game? *Drug Abuse and Alcoholism Newsletter, 8*, 4.

Comitas, L. (1976). Cannabis and work in Jamaica: A refutation of the amotivational syndrome. *Annals of the New York Academy of Sciences*, 282, 24–34.

Cone, E. J., Johnson, R. E., Paul, B. D., Mell, L. D., & Mitchell, J. (1988). Marijuana-laced brownies: Behavioral effects, physiologic effects, and urinalysis in humans following ingestion. *Journal of Analytic Toxicology, 12*, 169–175.

Consroe, P., Laguna, J., Allender, J., Snider, S., Stern, L., Sandyk, R., et al. (1991). Controlled clinical trial of cannabidiol in Huntington's disease. *Pharmacology, Biochemistry and Behavior, 40*, 701–708.

Consroe, P., Musty, R., Rein, J., Tillery, W., & Pertwee, R. G. (1997). The perceived effects of smoked cannabis on patients with multiple sclerosis. *European Neurology, 38*, 44–48.

Consroe, P., Sandyk, R., & Snider, S. R. (1986). Open label evaluation of cannabidiol in dystonic movement disorders. *International Journal of Neuroscience, 30*, 277–282.

Coombs, R. H., & West, L. J. (1991). Drug testing: Issues and options. New York: Oxford University Press.

Cornish, J. W., McNicholas, L. F., & O'Brien, C. P. (1995). Treatment of substance related disorders. In A. F. Schatzberg & C. B. Nemeroff (Eds.), *Textbook of psychopharmacology* (pp. 575–637). Washington, DC: American Psychiatric Association.

Cosgrove, J., & Newell, T. G. (1991). Recovery of neuropsychological functions during reduction in use of phencyclidine. *Journal of Clinical Psychology, 47*, 159–169.

Cousens, K., & DiMascio, A. (1973). Delta-9-THC as an hypnotic: An experimental study of three dose levels. *Psychopharmacologia, 33*, 355–364.

Crane v. Campbell, 245 U.S. 304; 38 S. Ct. 98 (1917).

Creason, C. R., & Goldman, M. (1981). Varying levels of marijuana use by adolescents and the amotivational syndrome. *Psychological Reports, 48*, 447–454.

Culver, C. M., & King, F. W. (1974). Neuropsychological assessment of undergraduate marihuana and LSD users. *Archives of General Psychiatry, 31,* 707–711.

Cunha, J. M., Carlini, E. A., Pereira, A. E., Ramos, O. L., Pimental, C., Gagliardi, R., et al. (1980). Chronic administration of cannabidiol to healthy volunteers and epileptic patients. *Pharmacology, 21,* 175–185.

da Orta, G. (1563/1913). *Colloquies on the simples & drugs of India (Goa).* (Sir Clements Markham, Trans.). London: Henru Sotheran.

Dansak, D. A. (1997). As an antiemetic and appetite stimulant for cancer patients. In M. L. Mathre (Ed.), *Cannabis in medical practice* (pp. 69–83). London: McFarland.

Davidson, E. S. & Schent, S. (1994). Variability in subjective responses to marijuana: Initial experiences of college students. *Addictive Behaviors, 19,* 531–538.

Davis, K. H., Jr., McDaniel, I. A., Jr., Cadwell, L. W., & Moody, P. L. (1984). Some smoking characteristics of marijuana cigarettes. In S. Agurell, W. L. Dewey, & R. E. Wilette (Eds.), *Cannabinoids: Chemical, pharmacologic, and therapeutic aspects* (pp. 97–110). New York: Academic.

Dawes, R. M. (1994). *House of cards.* New York: Free Press.

Day, N. L., Richardson, G. A., Geva, D., & Robles, N. (1994). Alcohol, marijuana, and tobacco: Effects of prenatal exposure on offspring growth and morphology at age six. *Alcoholism: Clinical and Experimental Research, 18,* 786–794.

De Petrocellis, L., Melck, D., Bisogno, T., Milone, A. & Di Marzo, V. (1999). Finding of the endocannabinoid signalling system in Hydra, a very primitive organism: Possible role in the feeding response. *Neuroscience, 92,* 377–387.

De Petrocellis, L., Melck, D., Palmisano, A., Bisogno, T., Laezza, C., Bifulco, M., et al. (1998). The endogenous cannabinoid anandamide inhibits human breast cancer cell proliferation. *Proceedings of the National Academy of Sciences, 95,* 8375–8380.

De Quincy, T. (1966). *Confessions of an opium eater.* New York: American Library.

de Zwart, W. M., Stam, H., & Kuiplers, S.B.M. (1997). *Key data—smoking, drinking, drug use, and gambling among pupils aged 10 years or older.* Netherlands: Netherlands Institute of Health and Addiction.

Dennis, R. J. (1990). The American people are starting to question the drug war. In A. S. Trebach & K. B. Zeese (Eds.), *The great issues in drug policy* (pp. 141–186). Washington, DC: Drug Policy Foundation.

Department of Health and Human Services (DHHS). (1998). *National household survey on drug abuse: Population estimates, 1997.* Washington, DC: U.S. Government Printing Office.

Devane, W. A., Dysarz, F. A., Johnson, M. R., Melvin, L. S., & Howlett, A. C. (1988). Determination and characterization of a cannabinoid receptor in rat brain. *Molecular Pharmacology, 34,* 605–613.

Devane, W. A., Hanus, L., Breuer, A., Pertwee, R. G., Stevenson, L. A., Griffin, G., et al. (1992). Isolation and structure of a brain constituent that binds to the cannabinoid receptor. *Science, 258,* 1946–1949.

Di Marzo, V., Sepe, N., De Petrocellis, L., Berger, A., Crozier, G., Fride, E., et al. (1998). Trick or treat from food cannabinoids? *Nature, 396,* 636.

Diaz, J. (1997). *How drugs influence behavior.* Upper Saddle River, NJ: Prentice Hall.

Dixon, L., Haas, G., Weiden, P. J., Sweeney, J., & Frances, A. J. (1991). Drug abuse in schizophrenic patients: Clinical correlates and reasons for use. *American Journal of Psychiatry, 148,* 224–230.

Doblin, R. (1994). The MAPS/California NORML marijuana waterpipe/vaporizer study. *Newsletter of the Multidisciplinary Association for Psychedelic Studies, 5,* 19–22.

Donaldson, S. I., Sussman, S., MacKinnon, D. P., Severson, H. H., Glynn, T., Murray, D. M., et al. (1996). Drug abuse prevention programming: Do we know what content works? *American Behavioral Scientist, 39,* 868–883.

Donovan, J. E. (1996). Problem behavior theory and the explanation of adolescent marijuana use. *Journal of Drug Issues, 26,* 379–404.

Donovan, J. E., & Jessor, R. (1983). Problem drinking and the dimension of involvement with drugs: A Guttman scalogram analysis of adolescent drug use. *American Journal of Public Health, 73,* 543–552.

Doorenbos N., Fetterman, P., Quimby, M., & Turner, C. (1971). Cultivation, extraction, and analysis of *Cannabis sativa L. Annals of the New York Academy of Sciences, 191,* 3–14.

Dornbush, R. L. (1974). Marijuana and memory: Effects of smoking on storage. *Transactions of the New York Academy of Science, 36,* 94–100.

Dornbush, R. L., Fink, M., & Freedman, A. M. (1971). Marijuana, memory, and perception. *American Journal of Psychiatry, 128,* 194–197.

Doweiko, H. E. (1999). *Concepts of chemical dependency.* New York: Brooks Cole.

Dreher, M. C. (1997). Cannabis and pregnancy. In M. L. Mathre (Ed.)., *Cannabis in medical practice* (pp. 159–170). London: McFarland.

Dreher, M. C., Nugent, K., & Hudgins, R. (1994). Prenatal marijuana exposure and neonatal outcomes in Jamaica: An ethnographic study. *Pediatrics, 93,* 254–260.

Drug Watch Oregon. (1996). *Marijuana research review.* Portland, OR: Author.

Drummer, O. H. (1994). Drugs in drivers killed in Australian road traffic accidents. (Report no. 0594). Melbourne, Australia: Monash University, Victorian Institute of Forensic Pathology.

Du Toit, B. M. (1975). Dagga: The history and ethnographic setting of *Cannabis sativa* in Southern Africa. In V. Rubin (Ed.), *Cannabis and culture* (pp. 51–62). The Hague: Mouton.

Du Toit, B. M. (1980). *Cannabis in Africa*. Rotterdam: Balkema.

Dumas, A. (1844/1998). *The Count of Monte Cristo*. New York: Oxford University Press.

Duncan, D. F. (1987). Lifetime prevalence of "amotivational syndrome" among users and non-users of hashish. *Psychology of Addictive Behaviors, 1,* 114–119.

Dunn, M., & Davis, R. (1974). The perceived effects of marijuana on spinal cord injured males. *Paraplegia, 12,* 175.

DuPont, R. (1984). *Getting tough on gateway drugs*. Washington, DC: American Psychiatric Association.

Dutch Ministry of Health, Welfare, and Sport. (1995). *Drug policy in the Netherlands—continuity and change*. Netherlands: Author.

Earleywine, M., & Finn, P. R. (1991). Sensation seeking explains the relation between behavioral inhibition and drinking. *Addictive Behaviors, 16,* 123–128.

Earleywine, M., Finn, P. R., & Martin C. S. (1990). Personality risk for alcoholism and alcohol consumption: A latent variable analysis. *Addictive Behaviors, 15,* 183–187.

Earleywine, M., & Newcomb, M. (1997). Concurrent versus simultaneous polydrug use: Prevalence, correlates, discriminant validity, and prospective effects on health outcomes. *Experimental and Clinical Psychopharmacology, 5,* 353–364.

Eaton, C. (1966). *A history of the old south*. New York: Macmillan.

Eddy, N. B., Halbach, H., Isbell, H., & Seevers, M. H. (1965). Drug dependence: Its significance and characteristics. *Bulletin of the World Health Organization, 32,* 721–733.

Elmore, A. M., & Tursky, B. (1981). A comparison of two psychophysiological approaches to the treatment of migraine. *Headache, 21,* 93–101.

ElSohly, M. A. Holley, J. H., & Turner, C. E. (1985). Constituents of *Cannabis sativa L.* XXVI. The delta-9-tetrahydrocannabinol content of confiscated marijuana, 1974–1983. In D. J. Harvey (Ed.), *Marijuana '84* (pp. 233–247). Oxford: IRL.

ElSohly, M. A., Ross, S. A., Mehmedic, Z., Arafat, R., Yi, B., & Banahan, B. F. (2000) Potency trends of delta-9-THC and other cannabinoids in confiscated marijuana from 1980–1997. *Journal of Forensic Sciences, 45,* 24–30.

Employment Division v. Smith, 494 U.S. 872 (1990).

Emrich, H. M., Weber, M. M., Wendl, A., Zihl, J., Von Meyer, L., & Hanishc, W. (1991). Reduced binocular depth inversion as an indicator of cannabis induced censorship impairment. *Pharmacology, Biochemistry and Behavior, 40,* 689–690.

Ennett, S. T., Tobler, N. S., Ringwalt, C. L., & Flewelling, R. L. (1994). How effective is drug abuse resistance education? A meta-analysis of Project DARE outcome evaluations. *American Journal of Public Health, 84,* 1394–1401.

Entin, E. E., & Goldzung, P. J. (1973). Residual effects of marijuana use on learning and memory. *Psychological Record, 23,* 169–178.

Evans, L. (1999). Last words: Wedding day dreams. *Hemp Times, 3,* 90.

Evans, M. A., Martz, R., Brown, D. J., Rodda, B. E., Kiplinger, G. F., Lemberger, L., et al. (1973). Impairment of performance with low doses of marihuana. *Clinical Pharmacology and Therapeutics, 14,* 936–940.

Evans, M. A., Martz, R., Rodda, B. E., Lemberger, L., & Forney, R. B. (1976). Effects of marihuana-dextroamphetamine combination. *Clinical Pharmacology and Therapeutics, 20,* 350–361.

Evans, M. D., Hollon, S. D., Derubeis, R. J., Pinsecki, J. M., Grove, W. M., Garvey, J. J., et al. (1992). Differential relapse following cognitive therapy and pharmacotherapy for depression. *Archives of General Psychiatry, 49,* 802–808.

Evans, R. M. (1998). What is "legalization"? What are "drugs"? In J. M. Fish (Ed.), *How to legalize drugs* (pp. 369–387). Northvale, NJ: Jason Aronson.

Federal Bureau of Investigation. (1997). *Crime in the United States, 1996, FBI uniform crime report.* Washington, DC: U.S. Government Printing Office.

Ferguson, T. J., Rule, B. G., & Lindsay, R. C. (1982). The effects of caffeine and provocation on aggression. *Journal of Research in Personality, 16,* 60–71.

Fiore, M. C., Smith, S. S., Jorenby, D. E., & Baker, T. B. (1994). The effectiveness of the nicotine patch for smoking cessation. *Journal of the American Medical Association, 271,* 1940–1947.

Fiorentine, R., & Anglin, M. D. (1997). Does increasing the opportunity for counseling increase the effectiveness of outpatient drug treatment? *American Journal of Drug and Alcohol Abuse, 23,* 369–382.

Fish, J. M. (Ed.). (1998). *How to legalize drugs.* Northvale, NJ: Jason Aronson.

Fletcher, J. M., & Satz, P. (1977). A methodological commentary on the Egyptian study of chronic hashish use. *Bulletin on Narcotics, 29,* 29–34.

Foltin, R. W., Fischman, M. W., Brady, J. V., Bernstein, D. J., Capriotti, R. M., Nellis, M. J., et al. (1990). Motivational effects of smoked marijuana: Behavioral contingencies and low-probability activities. *Journal of the Experimental Analysis of Behavior, 53,* 5–19.

Foltin, R. W., Fischman, M. W., Brady, J. V., Kelly, T. H., Bernstein, D. J., & Nellis, M. J. (1989). Motivational effects of smoked marijuana: Behavioral contingencies and high-probability recreational activities. *Pharmacology, Biochemistry and Behavior, 34,* 871–877.

Foltin, R. W., Fischman, M. W., & Byrne, M. F. (1988). Effects of smoked marijuana on food intake and body weight of humans living in a residential laboratory. *Appetite, 11,* 1–14.

Fossier, A. E. (1931). The marijuana menace. *New Orleans Medical and Surgical Journal, 84,* 247–252.

Foucault, M. (1973). *Madness and civilization: A history of insanity in the age of reason.* New York: Random House.

Frankel, J. P., Hughes, A., Lees, A. J., & Stern, G. M. (1990). Marijuana for parkinsonian tremor. *Journal of Neurology, Neurosurgery and Psychiatry, 53,* 436.

Franklin, D. (1990). Hooked-not hooked: Why isn't everyone an addict? *Health, 9,* 39–52.

Franzini, L. R., & Grossberg, J. M. (1995). *Eccentric and bizarre behaviors.* New York: Wiley.

Fride, E., & Mechoulam, R. (1993). Pharmacological activity of the cannabinoid receptor agonist, anandamide, a brain constituent. *European Journal of Pharmacology, 231,* 313–314.

Fried, P. A., Watkinson, B., & Gray, R. (1992). A follow-up study of attentional behavior in 6-year-old children exposed prenatally to marijuana, cigarettes, and alcohol. *Neurotoxicology and Teratology, 14,* 299–311.

Fried, P. A., Watkinson, B., & Willan, A. (1984). Marijuana use during pregnancy and decreased length of gestation. *American Journal of Obstetrics and Gynecology, 150,* 23–27.

Fuentes, J. A., Ruiz-Gayo, M., Manzanares, J., Vela, G., Reche, I., & Corchero, J. (1999). Cannabinoids as potential new analgesics. *Life Sciences, 65,* 675–685.

Garraty, J. A. & Gay, P. (1981). *The Columbia history of the world.* New York: Harper and Row.

Gautier, T. (1846/1966). The hashish club. In D. Solomon (Ed.), *The marijuana papers* (pp. 121–135). New York: Bobbs-Merrill.

Gerard, C. M., Mollereau, C., Vassart, G., & Parmentier, M. (1991). Molecular cloning of a human cannabinoid receptor which is also expressed in testis. *Biochemistry Journal, 279,* 129–134.

Gergen, M. K., Gergen, K. J., & Morse, S. J. (1972). Correlates of marijuana use among college students. *Journal of Applied Social Psychology, 2,* 1–16.

Gianutsos, R., & Litwack, A. R. (1976). Chronic marijuana smokers show reduced coding into long-term storage. *Bulletin of the Psychonomic Society, 7,* 277–279.

Gieringer, D. (1996). Marijuana water pipe and vaporizer study. *Newsletter of the Multidisciplinary Association for Psychedelic Studies, 6,* 5–9.

Ginsberg, A. (1966). First manifesto to end the bringdown. In D. Solomon (Ed.), *The marijuana papers* (pp. 183–200). New York: Bobbs-Merrill.

Godwin, H. (1967). The ancient cultivation of hemp. *Antiquity, 41,* 42–49.

Gold, D. (1989). *Cannabis alchemy: The art of modern hashmaking.* Berkeley: Ronin.

Goldberg, L., Bents, R., Bosworth, E., Trevistan, L., & Elliot, D. C. (1991). Anabolic steroid education and adolescents: Do scare tactics work? *Pediatrics, 87,* 283–286.

Goldberg, R. (1997). *Drugs across the spectrum.* Englewood, CO: Morton.

Goldschmidt, L., Day, N. L., & Richardson, G. A. (2000). Effects of prenatal marijuana exposure on child behavior problems at age 10. *Neurotoxicology and Teratology, 22,* 325–336.

Golub, A. & Johnson, B. D. (1994). The shifting importance of alcohol and marijuana as gateway substances among serious drug abusers. *Journal of Studies on Alcohol, 55,* 607–614.

Goode, E. (1971). Drug use and grades in college. *Nature, 239,* 225–227.

Gordon, D. R. (1994). *The return of the dangerous classes—drug prohibition and policy politics.* New York: W. W. Norton.

Gorenstein, E. E. (1987). Cognitive-perceptual deficits in an alcoholism spectrum disorder. *Journal of Studies on Alcohol, 48,* 310–318.

Gorenstein, E. E., Mammato, C. A., & Sandy, J. M. (1989). Performance of inattentive-overactive children on selected measures of prefrontal-type function. *Journal of Clinical Psychology, 45,* 619–632.

Gorman, T. J. (1996). *Marijuana is NOT medicine.* Santa Clarita, CA: California Narcotic Officers' Association.

Gorter, R. (1991). Management of anorexia-cachexia associated with cancer and HIV infection. *Oncology (Supplement), 5,* 13–17.

Gougeon, D. (1984–1985). CEEB SAT mathematics scores and their correlation with college performance in math. *Educational Research Quarterly, 9,* 8–11.

Gralla, R. J., Tyson, L. B., Bordin, L. A., Clark, R. A., Kelsen, D. P., Kris,

M. G., et al. (1984). Antiemetic therapy: A review of recent studies and a report of a random assignment trial comparing metoclopramide with delta-9-tetrahydrocannabinol. *Cancer Treatment Reports, 68,* 163–172.

Grant, B. F., & Pickering, R. (1998). The relationship between cannabis use and DSM-IV cannabis abuse and dependence: Results from the national longitudinal alcohol epidemiological survey. *Journal of Substance Abuse, 10,* 255–264.

Grant, I., Rochford, J., Fleming, T., & Stunkard, A. (1973). A neuropsychological assessment of the effects of moderate marihuana use. *Journal of Nervous and Mental Disease, 156,* 278–280.

Grattan, J. H. G., & Singer, C. (1952). *Anglo-Saxon magic and medicine.* London: Oxford University Press.

Gray, L. C. (1958). *History of agriculture in the Southern United States.* Gloucester, MA: Peter Smith.

Green, B. E., & Ritter, C. (2000). Marijuana use and depression. *Journal of Health and Social Behavior, 41,* 40–49.

Greenfield, S. F., & O'Leary, G. (1999). Sex differences in marijuana use in the United States. *Harvard Review of Psychiatry, 6,* 297–303.

Greenwald, M. K., & Stitzer, M. L. (2000). Antinoceptive, subjective and behavioral effects of smoked marijuana in humans. *Drug and Alcohol Dependence, 59,* 261–275.

Grigor, J. (1852). Indian hemp as an oxytocic. *Monthly Journal of Medical Science, 15,* 124–125.

Grilly, D. M. (1998). *Drugs and human behavior.* Boston: Allyn and Bacon.

Grinspoon, L. (1971). *Marijuana reconsidered.* Cambridge: Harvard University Press.

Grinspoon, L., & Bakalar, J. B. (1997). *Marijuana, the forbidden medicine.* New Haven: Yale University Press.

Gross, H., Egbert, M. H., Faden, V. B., Godberg, S. C., Kaye, W. H., Caine, E. D., et al. (1983). A double-blind trial of delta-9-THC in primary anorexia nervosa. *Journal of Clinical Psychopharmacology, 3,* 165–171.

Gruber, A. J., Pope, H. G., & Oliva, P. (1997). Very long-term users of marijuana in the United States: A pilot study. *Substance Use and Misuse, 32,* 249–264.

Grun, B. (1982). *The timetables of history.* New York: Touchstone.

Halikas, J. A., Goodwin, D. W., & Guze, S. B. (1971). Marijuana effects: A survey of regular users. *Journal of the American Medical Association, 217,* 692–694.

Halikas, J. A., Weller, R. A. & Morse, C. L. (1982). Effects of regular marijuana use on sexual performance. *Journal of Psychoactive Drugs, 14,* 59–70.

Halikas, J. A., Weller, R. A., Morse, C. L., & Hoffmann, R. G. (1985). A longitudinal study of marijuana effects. *International Journal of the Addictions, 20,* 701–711.

Hall, W. & Solowij, N. (1998). Adverse effects of cannabis. *Lancet, 352,* 1611–1616.

Hall, W., Solowij, N., & Lennon, J. (1994). *The health and psychological consequences of cannabis use.* Canberra: Australian Government Publication Services.

Haney, M., Ward, A. S., Comer, S. D., Foltin, R. W., & Fischman, M. W. (1999a). Abstinence symptoms following oral THC administration to humans. *Psychopharmacology, 141,* 385–394.

Haney, M., Ward, A. S., Comer, S. D., Foltin, R. W., & Fischman, M. W. (1999b). Abstinence symptoms following smoked marijuana in humans. *Psychopharmacology, 141,* 395–404.

Hanigan, W. C., Destree, R., & Truong, X. T. (1986). The effect of delta-9-THC on human spasticity. *Clinical Pharmacology and Therapeutics, 39,* 198.

Hannerz, J., & Hindmarsh, T. (1983). Neurological and neuroradiological examination of chronic cannabis smokers. *Annals of Neurology, 13,* 207–210.

Hansen, W. B. (1992). School-based substance abuse prevention: A review of the state of the art in curriculum, 1980–1990. *Health Education Research: Theory and Practice, 7,* 403–430.

Harris, L. S., Munson, A. E., & Carchman, R. A. (1976). Antitumor properties of cannabinoids. In M. C. Braude and S. Szara (Eds.), *The pharmacology of marijuana.* (Vol. 2, pp. 773–776). New York: Raven.

Hartley, J. P., Nogrady, S. G., & Seaton, A. (1978). Bronchodilator effect of delta-1-tetrahydrocannabinol. *British Journal of Clinical Pharmacology, 5,* 523–525.

Hasan, K. A. (1974). Social aspects of the use of cannabis in India. In V. Rubin (Ed.), *Cannabis and culture* (pp. 235–246). The Hague: Mouton.

Hayes, J. S., Lampart, R., Dreher, M. C., & Morgan, L. (1991). Five-year follow-up of rural Jamaican children whose mothers used marijuana during pregnancy. *West Indian Medical Journal, 40,* 120–123.

Heishman, S. J., Huestis, M. A., Henningfield, J. E., & Cone, E. J. (1990). Acute and residual effects of marijuana: Profiles of plasma THC levels, physiological, subjective and performance measures. *Pharmacology, Biochemistry and Behavior, 34,* 561–565.

Heishman, S. J., Stitzer, M. L., & Yingling, J. E. (1989). Effects of tetrahydrocannabinol content on marijuana smoking behavior, subjective reports, and performance. *Pharmacology, Biochemistry & Behavior, 34,* 173–179.

Hembree, W. C., Nahas, G. G., Zeidenberg, P., & Huang, H. F. S. (1979). Changes in human spermatozoa associated with high-dose marijuana smoking. In G. G. Nahas & W. D. M. Paton (Eds.), *Marijuana: Biological effects, analysis, metabolism, cellular responses* (pp. 429–439). New York: Pergamon.

Hemming, M., & Yellowlees, P. M. (1993). Effective treatment of Tourette's syndrome with marijuana. *Journal of Psychopharmacology*, 7, 389–391.

Henningfield, J., Cohen, C., & Pickworth, W. (1993). Psychopharmacology of nicotine. In C. Orleans & J. Slade (Eds.), *Nicotine addiction: Principle and management* (pp. 24–45). New York: Oxford University Press.

Hepler, R. S., & Petrus, R. (1971). Experiences with administrations of marijuana to glaucoma patients. In S. Cohen and R. Stillman (Eds.), *The therapeutic potential of marijuana* (pp. 63–76). New York: Plenum.

Herer, J. (1999). *The emperor wears no clothes*. Van Nuys, CA: HEMP Publishing.

Herkenham, M., Lynn, A. B., Little, M. D., Johnson, M. R., Melvin, L. S., De Costa, B. R., et al. (1990). Cannabinoid receptor localization in brain. *Proceedings of the National Academy of Sciences, 87,* 1932–1936.

Herodotus (1999/5th Century B.C.). *The histories*. C. Dewald (Ed.). (R. A. Waterfield, Trans.). New York: Oxford University Press.

Hill, S. Y., Schwin, R., Goodwin, D. W., & Powell, B. J. (1974). Marijuana and pain. *Journal of Pharmacology and Experimental Therapeutics, 188,* 415–418.

Hilts, P. J. (1994, August 2). Is nicotine addictive? It depends on whose criteria you use. *New York Times*, p. C3.

Himmelstein, J. L. (1986). The continuing career of marijuana: Backlash . . . within limits. *Contemporary Drug Problems, 13,* 1–21.

Hochman, J. S., & Brill, N. Q. (1973). Chronic marijuana use and psychosocial adaptation. *American Journal of Psychiatry, 130,* 132–139.

Hoefler, M., Lieb, R., Perkonigg, A., Schuster, P., Sonntag, H., & Wittchen, H. U. (1999). Covariates of cannabis use progression in a representative population sample of adolescents: A prospective examination of vulnerability and risk factors. *Addiction, 94,* 1679–1694.

Hollister, L. E. (1974). Structure-activity relationships in man of cannabis constituents, and homologs and metabolites of delta-9 tetrahydrocannabinol. *Pharmacology, 11,* 3–11.

Holloway, M. (1991). Rx for addiction. *Scientific American, 264,* 94–103.

Holly, E. A., Lele, C., Bracci, P. M., & McGrath, M. S. (1999). Case-control study of non-Hodgkin's lymphoma among women and heterosexual men in the San Francisco Bay area. *American Journal of Epidemiology, 150,* 375–389.

Hooker, W. D., & Jones, R. T. (1987). Increased susceptibility to memory intrusions and the Stroop interference effect during acute mairjuana intoxication. *Psychopharmacology, 91,* 20–24.

Hope, D. A., & Heimberg, R. G. (1993). Social phobia and social anxiety. In D. Barlow (Ed.), *Clinical handbook of psychological disorders* (pp. 99–136). New York: Guilford.

Horton, J. P., Nogrady, S. G., & Seaton, A. (1978). Bronchodilator effect of delta-1-tetrahydrocannabinol. *British Journal of Clinical Pharmacology, 5,* 523–525.

House of Lords—Select Committee on Science and Technology (1998). *Cannabis—The scientific and medical evidence.* London: The Stationery Office.

How much marijuana do Americans really smoke? (1995). *Forensic Drug Abuse Advisor, 7,* 7–8.

Howlett, A. C., Evans, D. M., & Houston, D. B. (1992). The cannabinoid receptor. In L. Murphy & A. Bartke (Eds.), *Marijuana/cannabinoids: Neurobiology and neurophysiology* (pp. 387–423). Boca Raton, FL: CRC.

Howlett, A. C., Johnson, M. R., Melvin, L. S., & Milne, G. M. (1988). Nonclassical cannabinoid analgesics inhibit adenylate cyclase: Development of a cannabinoid receptor model. *Molecular Pharmacology, 33,* 297–302.

Hser, Y. I., Grella, C., Chou, C. P., & Anglin, M. D. (1998). Relationships between drug treatment careers and outcomes: Findings from the National Drug Abuse Treatment Outcome Study. *Evaluation Review, 22,* 496–519.

Huestis, M. A., & Cone, E. J. (1998). Urinary excretion half-life of 11-Nor-9-carboxy-DELTA-9-tetrahydrocannabinol in humans. *Proceedings of the Fifth International Congress of Therapeutic Drug Monitoring and Clinical Toxicology, 20,* 570–576.

Hume, D. (1739/1978). *A treatise on human nature.* New York: Oxford University Press.

Hunt, C. A., & Jones, R. T. (1980). Tolerance and disposition of tetrahydrocannabinol in man. *Journal of Pharmacology and Experimental Therapeutics, 215,* 35–44.

Hunt, W. A., Barnett, L. W., & Branch, L. G. (1971). Relapse rates in addiction programs. *Journal of Clinical Psychology, 27,* 455–456.

Husak, D. (1992). *Drugs and rights.* New York: Cambridge University Press.

Husak, D. (1998). Two rationales for drug policy: How they shape the content of reform. In J. M. Fish (Ed.), *How to legalize drugs* (pp. 29–60). Northvale, NJ: Jason Aronson.

Hymowitz, N., Feuerman, J., Hollander, M., & Frances, R. J. (1993). Smoking deterrence using silver acetate. *Hospital and Community Psychiatry, 44,* 113–116.

Indian Hemp Drugs Commission (IHDC). (1894). Report of the Indian hemp drugs commission. Simla, India: Government Central Printing Office.

Indiana Prevention Resource Center. (1998). *Factline on marijuana.* Bloomington: The Trustees of Indiana University.

Institute of Medicine (IOM). (1999). *Marijuana and medicine: Assessing the science base.* Washington, DC: National Academy.

Iversen, L. L. (2000). *The science of marijuana.* New York: Oxford University Press.

Jain, A. K., Ryan, J. R., McMahon, F. G., & Smith, G. (1981). Evaluation of intramuscular levonantradol and placebo in acute postoperative pain. *Journal of Clinical Pharmacology, 21,* 320S–326S.

Jarai, Z., Wagner, J. A., Goparaju, S. K., Wang, L., Razdan, R. K., Sugiura, T., et al. (2000). Cardiovascular effects of 2-AG in anesthetized mice. *Hypertension, 35,* 679–684.

Jarbe, T. U., & Hiltunen, A. J. (1987). Cannabimimetic activity of cannabinol in rats and pigeons. *Neuropharmacology, 26,* 219–228.

Jessor, R. (1998). *New perspective on adolescent risk behaviors.* Cambridge: Cambridge University Press.

Jessor, R., & Jessor, S. L. (1977). *Problem behavior and psychosocial development: A longitudinal study of youth.* New York: Academic Press.

Joesof, M. R., Beral, V., Aral, S. O., Rolfs, R. T., & Cramer, D. W. (1993). Fertility and use of cigarettes, alcohol, marijuana, and cocaine. *Annals of Epidemiology, 3,* 592–594.

Johansson, E., Arguell, S., Hollister, L., & Halldin, M. (1988). Prolonged apparent half-life of delta-1-tetrahydrocannabinol in plasma of chronic marijuana users. *Journal of Pharmacy and Pharmacology, 40,* 374–375.

Johansson, E., Noren, K., Sjovall, J., & Halldin, M. M. (1989). Determination of delta-1-tetrahydrocannabinol in human fat biopsies from marihuana users by gas chromatography-mass spectrometry. *Biomedical Chromatography, 3,* 35–38.

Johnston, J. F. (1855). *Chemistry of common life.* New York: Appleton.

Johnston, L., Bachman, J., & O'Malley, P. (1981). Marijuana decriminalization: The impact on you, 1975–1980. *Monitoring the Future Occasional Paper, 13,* 27–29.

Johnston, L., Bachman, J., & O'Malley, P. (1996). *National survey results on drug use from the Monitoring the Future study, 1975–1995.* Washington, DC: U.S. Government Printing Office.

Kabilek, J., Krejci, Z., & Santavy, F. (1960). Hemp as a medicament. *Bulletin on Narcotics, 12,* 5–22.

Kaestner, R. (1991). The effects of drug use on the wages of young adults. *Journal of Labor Economics, 9,* 381–412.

Kaestner, R. (1994a). The effect of illicit drug use on the labor supply of young adults. *Journal of Human Resources, 29,* 123–136.

Kaestner, R. (1994b). New estimates of the effect of marijuana and cocaine on wages: Accounting for unobserved person specific effects. *Industrial and Labor Relations Review, 47,* 454–470.

Kandel, D. B., & Davies, M. (1992). Progression to regular marijuana involvement: Phenomenology and risk factors for near-daily use. In M. Glantz & R. Pickens (Eds.), *Vulnerability to drug abuse* (pp. 211–253). Washington, DC: American Psychological Association.

Kandel, D. B., & Davies, M. (1996). High school students who use crack and other drugs. *Archives of General Psychiatry, 53,* 71–80.

Kandel, D. B., Yamaguchi, K., & Chen, K. (1992). Stages of progression in drug involvement from adolescence to adulthood: Further evidence for the gateway theory. *Journal of Studies on Alcohol, 53,* 447–457.

Kaplan, J. (1970). *Marijuana—The new prohibition.* New York: Wald.

Karacan, D. B., Fernandez-Salas, A., Coggins, W. J., Carter, W. E., Williams, R. L., Thornby, J. I., et al. (1976). Sleep electroencephalagraphic–electrooculographic characteristics of chronic marijuana users. *Annals of the New York Academy of Sciences, 282,* 348–374.

Karniol, I. G., Shirakawa, I., Takahashi, R. N., Knobel, E., & Musty, R. E. (1975). Effects of delta-9-tetrahydrocannabinol and cannabinol in man. *Pharmacology, 13,* 502–512.

Kaslow, R. A., Blackwelder, W. C., Ostrow, D. G., Yerg, D., Palenicek, J., Coulson, A. H., et al. (1989). No evidence for a role of alcohol or other psychoactive drugs in accelerationg immunodeficiency in HIV-1-positive individuals. *Journal of the American Medical Association, 261,* 3424–3429.

Kattlove, H. (1995). Antiemetic properties of granisetron. *New England Journal of Medicine, 332,* 1653.

Kirk, D. (1999). From Hungary with love. *Hemp Times, 3,* 44–88.

Kirk, J. M., Doty, P., & de Wit, H. (1998). Effects of expectancies on subjective responses to oral delta-9-tetrahydrocannabinol. *Pharmacology, Biochemistry and Behavior, 59,* 287–293.

Kleiman, M.A.R. (1992). *Against excess: Drug policy for results.* New York: Basic Books.

Kleinhenz, J., Streitberger, K., Windeler, J., Gussbacher, A., Mavridis, G., & Martin, E. (1999). Randomised clinical trial comparing the effects of acupuncture and a newly designed placebo needle in rotator cuff tendinitis. *Pain, 83,* 235–241.

Koob, G. F., & Le Moal, M. (1997). Drug abuse: Hedonic homeostatic dysregulation. *Science, 278,* 52–58.

Koski, P. R., & Eckberg, D. L. (1983). Bureaucratic legitimation: Marihuana and the Drug Enforcement Administration. *Sociological Focus, 16,* 255–273.

Kotler, D. P., Tierney, A. R., Wang, J., & Pierson, R. N. (1989). Magnitude of body-cell-mass depletion and the timing of death from wasting in AIDS. *American Journal of Clinical Nutrition, 53,* 149–154.

Kouri, E., Pope, H. G., & Lukas, S. E. (1999). Changes in aggressive behavior during withdrawal from long-term marijuana use. *Psychopharmacology, 143,* 302–308.

Kouri, E., Pope, H. G., Yurgelun-Todd, D., & Gruber, S. (1995). Attributes of heavy vs. occasional marijuana smokers in a college population. *Biological Psychiatry, 38,* 475–481.

Krampf, W. (1997). AIDS and the wasting syndrome. In M. L. Mathre (Ed.), *Cannabis in medical practice* (pp. 84–93). London: McFarland.

Kuehnle, J., Mendelson, J. H., & David, K. R. (1977). Computed tomographic examination of heavy marijuana users. *Journal of the American Medical Association, 237,* 1231–1232.

Kung, C. T. (1959). *Archaeology in China.* Toronto: University of Toronto Press.

Kutchins, H., & Kirk, S. A. (1997). *Making us crazy.* New York: Free Press.

Labouvie, E., Bates, M. E., & Pandina, R. J. (1997). Age of first use: Its reliability and predictive utility. *Journal of Studies on Alcohol, 58,* 638–643.

LaBrie, J., & Earleywine, M. (2000). Sexual risk behaviors and alcohol: Higher base rate estimates revealed using the unmatched count technique. *Journal of Sex Research, 37,* 321–326.

Laird-Clowes, W. (1877). An amateur assassin. *Belgravia, 31,* 353–359.

Lapey, J. D. (1996). *Marijuana update 1996.* Omaha: Drug Watch International.

Lapp, W. M., Collins, R. L., Zywiak, W. H., & Izzo, C. V. (1994). Psychopharmacological effects of alcohol on time perception: The extended balanded placebo design. *Journal of Studies on Alcohol, 55,* 96–112.

Law, B., Mason, P. A., Moffat, A. C., Gleadle, R. I., & King, L. J. (1984). Forensic aspects of the metabolism and excretion of cannabinoids following oral ingestion of cannabis resin. *Journal of Pharmacy and Pharmacology, 36,* 289–294.

Leary v. U.S., 383 F.2d 851 (5th Cir. 1967).

Leary, T. (1997). *Flashbacks: A personal and cultural history of an era.* Los Angeles: J. P. Tarcher.

Leirer, V. O., Yesavage, J. A., & Morrow, D. G. (1991). Marijuana carry-over effects on aircraft pilot performance. *Aviation Space and Environmental Medicine, 62,* 221–227.

Lemberger, L., Axelrod, J., & Kopin, I. J. (1971). Metabolism and disposition of tetrahydrocannabinol in naive subjects and chronic marijuana users. *Pharmacological Reviews, 23,* 371–380.

Lemberger, L., & Rowe, H. (1975). Clinical pharmacology of nabilone, a cannabinol derivative. *Clinical Pharmacology & Therapeutics, 18,* 720–726.

Lenson, D. (1995). *On drugs.* Minneapolis: University of Minnesota Press.

Leuchtenberger, C. (1983). Effects of marijuana (cannabis) smoke on cellular biochemistry on In Vitro test systems. In K. O. Fehr and K. Kalant (Eds.), *Cannabis and health hazards* (pp. 177–224). Toronto: Addiction Research Foundation.

Levey, M. (1966). Medieval Arabic toxicology. *Transactions of the American Philosophical Society, 56,* 5–43.

Levinthal, C. (1999). *Drugs, behavior, and modern society.* Boston: Allyn and Bacon.

Leweke, F. M., Giuffrida, A., Wurster, U., Emrich, H. M., & Piomelli, D. (1999). Elevated endogenous cannabinoids in schizophrenia. *NeuroReport, 10,* 1665–1669.

Li, H. L. (1974). An archeological and historical account of cannabis in China. *Economic Botany, 28,* 437–448.

Li, H. L. (1975). The origin and use of cannabis in Eastern Asia: Their linguistic-cultural implications. In V. Rubin (Ed.), *Cannabis and culture* (pp. 51–62). The Hague: Mouton.

Light, G. A., Geyer, M. A., Clementz, B. A., Cadenhead, K. S., & Braff, D. L. (2000). Normal P50 suppression in schizophrenia patients treated with atypical antipsychotic medications. *American Journal of Psychiatry, 157,* 767–771.

Linn, S., Schoenbaum, S. C., Monson, R. R., Rosner, R., Stubblefield, P. C., & Ryan, K. J. (1983). The association of marijuana use with outcome of pregnancy. *American Journal of Public Health, 73,* 1161–1164.

Linzen, D. H., Dingemans, P. M., & Lenior, M. E. (1994). Cannabis abuse and the course of recent-onset schizophrenic disorders. *Archives of General Psychiatry, 51,* 273–279.

Low, M. D., Klonoff, H., & Marcus, A. (1973). The neurophysiological basis of the marijuana experience. *Canadian Medical Association Journal, 108,* 157–165.

Mann, P. (1985). *Marijuana alert.* New York: McGraw-Hill.

Manno, J. E., Kiplinger, G. F., Haine, S. E., Bennett, I. F., & Forney, R. B. (1970). Comparative effects of smoking marihuana or placebo on human motor and mental performance. *Clinical Pharmacology and Therapeutics, 11,* 808–815.

Margolin, B. (1998). *Guide to state and federal marijuana laws.* Los Angeles: Chuck Alton.

Marijuana Anonymous (1995). *Life with hope.* Van Nuys, CA: Marijuana Anonymous World Services.

Marlatt, A. (Ed.). (1998). *Harm reduction.* New York: Guilford.

Marlatt, G. A., Demming, B., & Reid, J. B. (1973). Loss of control drinking in alcoholics: An experimental analogue. *Journal of Abnormal Psychology, 81,* 233–241.

Marlatt, G. A., & Gordon, J. R. (1985). *Relapse prevention: Maintenance strategies in the treatment of addictive behaviors.* New York: Guilford.

Marlatt, G. A., & Rohsenow, D. J. (1980). Cognitive process in alcohol use: Expectancy and the balanced placebo design. In N. K. Mello (Ed.), *Advances in substance abuse.* (Vol. 1, pp. 159–199). Greenwich, CT: JAI.

Mathew, R. J., & Wilson, W. H. (1992). The effects of marijuana on cerebral blood flow and metabolism. In L. Murphy & A. Bartke (Eds.), *Marijuana/cannabinoids: Neurobiology and Neuro-physiology* (pp. 337–386). Boca Raton FL: CRC.

Mathew, R. J., Wilson, W. H., Chiu, N. Y., Turkington, T. G., Degrado, T. R., & Coleman, R. E. (1999). Regional cerebral blood flow and depersonalization after tetrahydrocannabinol administration. *Acta Psychiatrica Scandinavica, 100,* 67–75.

Mattes, R. D., Engelman, K., Shaw, L. M., & ElSohly, M. A. (1994). Cannabinoids and appetite stimulation. *Pharmacology, Biochemistry and Behavior, 49,* 187–195.

Mattes, R. D., Shaw, L. M., Edling-Owens, J., Engelman, K., & ElSohly, M. A. (1993). Bypassing the first-pass effect for the therapeutic use of cannabinoids. *Pharmacology, Biochemistry and Behavior, 44,* 745–747.

Mattes, R. D., Shaw, L. M., & Engelman, K. (1994). Effects of cannabinoids (marijuana) on taste intensity and hedonic ratings and salivary flow of adults. *Chemical Senses, 19,* 125–140.

Matthias, P., Tashkin, D. P., Marques-Magallanes, J. A., Wilkins, J. N., & Simmons, M. S. (1997). Effects of varying marijuana potency on deposition of tar and delta9-THC in the lung during smoking. *Pharmacology, Biochemistry and Behavior, 58,* 1145–1150.

Mc Daniel, M. A. (1988). Does pre-employment drug use predict job suita-

Ludlow, F. H. (1857). *The hasheesh eater: Being passages thagorean*. New York: Harper.

Lukas, S. E., Mendelson, J. H., & Benedikt, R. (1995). El graphic correlates of marijuana-induced euphoria. *Dru pendence, 37*, 131–140.

Lyketsos, C. G., Garrett, E., Liang, K. Y., & Anthony, J. bis use and cognitive decline in persons under 65 year *Journal of Epidemiology, 149*, 794–800.

Lynam, D. R., Milich, R., Zimmerman, R., Novak, S. P., tin, C., et al. (1999). Project DARE: No effects at 10 *Journal of Consulting and Clinical Psychology, 67*, 590–

Lyons, M. J., Toomey, R., Meyer, J. M., Green, A. I., Eis J., et al. (1997). How do genes influence marijuana us jective effects. *Addiction, 92*, 409–417.

Maccannell, K., Milstein, S. L., Karr, G., & Clark, S. (19 produced impairments in form perception: Experience experienced subjects. *Progress in Neuro-Psychopharmac*

MacCoun, R. J. (1993). Drugs and the law: A psychologi prohibition. *Psychological Bulletin, 113*, 497–512.

MacCoun, R., & Reuter, P. (1997). Interpreting Dutch ca soning by analogy in the legalization debate. *Science, 2*

MacDonald, D. I. (1984). *Drugs, drinking, and adolescents* Book Medical Publishers.

Mackesy-Amiti, M. E., Fendrich, M., & Goldstein, P. J. (1 drug use among serious drug users: Typical vs. atypica *and Alcohol Dependence, 45*, 185–196.

Mahdi, M. (Ed.). (1992). *The Arabian nights*. New York:

Maisto, S. A., Galizio, M., & Connors, G. J. (1995). *Drug* New York: Harcourt.

Makriyannis, A., & Rapaka, R. S. (1990). The molecular noid activity. *Life Sciences, 47*, 2173–2184.

Malec, J., Harvey, R. F., & Cayner, J. J. (1982). Cannabis in spinal cord injury. *Archives of Physical Medicine and* 116–118.

Maloff, D. (1981). A review of the effects of the decrimi juana. *Contemporary Drug Problems, 10*, 306–340.

Maltby, L. L. (1999). *Drug testing: A bad investment*. New Civil Liberties Union.

Mandel, J. (1988). Is marijuana law enforcement racist? *J tive Drugs, 20*, 83–91.

bility? In S. Guse & J. Welsh (Eds.), *Drugs in the workplace* (monograph 91, pp. 151–167). Rockville, MD: National Institute on Drug Abuse.

McGee, R., Williams, S. A., Poulton, R., & Moffitt, T. (2000). A longitudinal study of cannabis use and mental health from adolescence to early adulthood. *Addiction, 95,* 491–503.

McGeorge, J., & Aitken, C. K. (1997). Effects of cannabis decriminalization in the Australian Capital Territory on university students' patterns of use. *Journal of Drug Issues, 27,* 785–793.

McGlothlin, H. W., & West, L. J. (1968). The marijuana problem: An overview. *American Journal of Psychiatry, 125,* 1126–1134.

McKim, W. A. (1997). *Drugs and behavior: An introduction to behavioral pharmacology I.* Englewood Cliffs, NJ: Prentice Hall.

McMeens, R. R. (1860). Report of the committee on *Cannabis Indica.* In *Transactions of the 15th Annual Meeting of the Ohio State Medical Society.* Columbus, OH: Follett, Foster & Co. (Reprinted from *Marijuana: Medical papers, 1839–1972,* pp. 117–140, by T. H. Mikuriya, Ed., 1973, Oakland: Medi-Comp.)

McQuay, H., Carroll, D., & Moore, A. (1995). Variation in the placebo effect in randomized controlled trials of analgesics: All is as blind as it seems. *Pain, 64,* 331–335.

Mechoulam, R., Fride, E., Hanus, L., Sheskin, T., Bisogno, T., Di Marzo, V., et al. (1997). Anandamide may mediate sleep induction. *Nature, 389,* 25–26.

Meinck, H. M., Schonle, P. W., & Conrad, B. (1989). Effect of cannabinoids on spasticity and ataxia in multiple sclerosis. *Journal of Neurology, 236,* 120–122.

Mellaart, J. (1967). *Catal Huyuk: A neolithic town in Anatolia.* New York: McGraw-Hill.

Menhiratta, S. S., Wig, N. N., & Verma, S. K. (1978). Some psychological correlates of long-term heavy cannabis users. *British Journal of Psychiatry, 132,* 482–486.

Mikulas, W. L. (1996). Sudden onset of subjective dimensionality: A case study. *Perceptual and Motor Skills, 82,* 852–854.

Mikuriya, T. H. & Aldrich, M. R. (1988). Cannabis 1988: Old drug, new dangers, the potency question. *Journal of Psychoactive Drugs, 20,* 47–55.

Miles, C. G., Congreve, G. R. S., Gibbins, R. J., Marshman, J., Devenyi, P., & Hicks, R. C. (1974). An experimental study of the effects of daily cannabis smoking on behavior patterns. *Acta Pharmacologica et Toxicologica, 34*(Suppl. 7), 1–43.

Miller, C. & Wirtshafter, D. (1991). *The hemp seed cookbook.* Athens, OH: Hempery.

Miller, D. S., & Miller, T. Q. (1997). A test of socioeconomic status as a predictor of initial marijuana use. *Addictive Behaviors, 22,* 479–489.

Miller, L., & Cornett, T. (1978). Marijuana: Dose-response effects on pulse rate, subjective estimates of intoxication, free recall and recognition memory. *Pharmacology, Biochemistry and Behavior, 9,* 573–579.

Miller, L., Cornett, T., Drew, W., McFarland, D., Brightwell, D., & Wikler, A. (1977). Marijuana: Dose-response effects on pulse rate, subjective estimates of potency, pleasantness, and recognition memory. *Pharmacology, 15,* 268–275.

Miller, L., Cornett, T., & Wikler, A. (1979). Marijuana: Dose-response effects on pulse rate, subjective estimates of intoxication and multiple measures of memory. *Life Sciences, 25,* 1325–1350.

Miller, N. S., Gold, M. S., & Pottash, C. (1989). A 12-step treatment approach for marijuana (cannabis) dependence. *Journal of Substance Abuse Treatment, 6,* 241–250.

Miller, N. S., Gold, M. S., & Smith, D. E. (1997). *Manual of therapeutics for addictions.* New York: Wiley.

Miller, T. Q. (1994). A test of alternative explanations for the stage-like progression of adolescent substance use in four national samples. *Addictive Behaviors, 19,* 287–293.

Miller, W. R. (1999). *Integrating spirituality into treatment: Resources for practitioners.* Washington DC: American Psychological Association.

Miller, W. R., & Hester, R. K. (1986). Inpatient alcoholism treatment: Who benefits? *American Psychologist, 41,* 794–805.

Miller, W. R., & Rollnick, S. (1991). *Motivational interviewing.* New York: Guilford.

Miron, J. A. (1999). *Violence and the U.S. prohibition of drugs and alcohol.* (NBER Working Paper No. 6950. JEL No. K42).

Moeller, G. F., Dougherty, D. M., Lane, S. D., Steinberg, J. L., & Cherek, D. R. (1998). Antisocial personality disorder and alcohol-induced aggression. *Alcoholism: Clinical and Experimental Research, 22,* 1898–1902.

Molnar, J., Szabo, D., Pusztai, R., Mucsi, I., Berek, L., Ocsovszki, I., et al. (2000). Membrane associated antitumor effects of crocine-, ginsenoside, and cannabinoid derivatives. *Anticancer Research, 20,* 861–867.

Moos, R. H., King, M. J., & Patterson, M. A. (1996). Outcomes of residential treatment of substance abuse in hospital and community-based programs. *Psychiatric Services, 46,* 66–72.

Moreau, J. J. (1845/1973) *Hashish and mental illness.* New York: Raven.

Morganstern, J., Labouvie, E., McCrady, B. S., Kahler, C. W., & Frey, R. M. (1997). Affiliation with alcoholics anonymous after treatment: A study of its therapeutic effects and mechanisms of action. *Journal of Consulting and Clinical Psychology, 65,* 768–777.

Morley, S. (1997). Pain management. In A. Baum, S. Newman, J. Weinman, R. West, & C. McManus (Eds.), *Cambridge handbook of psychology, health, and medicine* (pp. 234–237). Cambridge: Cambridge University Press.

Morningstar, P. J. (1985). Thandai and Chilam: Traditional Hindu beliefs about the proper use of cannabis. *Journal of Psychoactive Drugs, 17,* 141–165.

Mueller, B. A., Daling, J. R., Weiss, N. S., & Moore, D. E. (1990). Recreational drug use and the risk of primary infertility. *Epidemiology, 1,* 195–200.

Muller-Vahl, K. R., Kolbe, H., & Dengler, R. (1997). Gilles de la Tourette syndrome: Influence of nicotine, alcohol, and marijuana on the clinical symptoms. *Der Nervenarzt, 68,* 985–989.

Musto, D. F. (1999). *The American disease: Origins of narcotic control.* New York: Oxford University Press.

Musty, R. E., & Kaback, L. (1995). Relationships between motivation and depression in chronic marijauna users. *Life Sciences, 56,* 2151–2158.

Myerscough, R., & Taylor, S. (1985). The effects of marijuana on human physical aggression. *Journal of Personality and Social Psychology, 49,* 1541–1546.

Nadelmann, E. A. (1992). Thinking seriously about alternatives to drug prohibition. *Daedalus, 121,* 87–132.

Nahas, G. G. (1986). Cannabis: Toxicological properties and epidemiological aspects. *Medical Journal of Australia, 145,* 82–87.

Nahas, G. G. (1990). *Keep off the grass.* Middlebury, VT: Paul S. Erickson.

Nahas, G. G., Suciv-Foca, G., Armand, J-P., & Morishima, A. (1974). Inhibition of cellular mediated immunity in marihuana smokers. *Science, 183,* 419–420.

Nakamura, E. M., da Sikes, E. A., Concho, G. K., Wilkinson, D. A., & Masur, J. (1991). Reversible effects of acute and long-term administration of delta 9 THC on memory in the rat. *Drugs and Alcohol Dependence, 28,* 167–175.

Narcotics Anonymous. (1988). *Narcotics anonymous.* Van Nuys, CA: World Services Office.

Nathan, P. (1988). The addictive personality is the behavior of the addict. *Journal of Consulting and Clinical Psychology, 56,* 183–188.

National Drug Strategy Household Survey Report (1995). Canberra: Australian Government Publishing Service.

National Institute on Drug Abuse (NIDA). (1991). *NIDA capsules: Summary of findings from the 1990 Household Survey on Drug Abuse*. Rockville, MD: U.S. Department of Health and Human Services.

National Institute on Drug Abuse (NIDA). (1997). *Monitoring the future study*. Washington, DC: U.S. Department of Health and Human Services. [December 20th press release].

National Institute on Drug Abuse (NIDA). (1998). *Marijuana: Facts parents need to know*. Washington, DC: U.S. Department of Health and Human Services.

National Organization for the Reform of Marijuana Laws (NORML). (1996a). Principles of responsible cannabis use. Available: http://www.natlnorml.org/about/responsible.shtml.

National Organization for the Reform of Marijuana Laws (NORML). (1996b). Crop earnings in the United States. Available: www.norml.org / facts /crop /report.shtml #croprank.

Needham, J. (1974). *Science and civilization in China*. Cambridge: Cambridge University Press.

Newcomb, M. & Earleywine, M. (1996). The willing host: Intrapersonal contributors to substance abuse. *American Behavioral Scientist, 7*, 823–837.

Newcomb, M. D., McCarthy, W. J., & Bentler, P. M. (1989). Cigarette smoking, academic lifestyle, and self-efficacy: An eight-year study from early adolescence to young adulthood. *Journal of Applied Social Psychology, 19*, 251–281.

Neylan, T. C., Fletcher, D. J., Lenoci, M., McCallin, K., Weiss, D. S., Schoenfeld, F. B., et al. (1997). Sensory gating in chronic post-traumatic stress disorder: Reduced auditory P50 suppression in combat veterans. *Biological Psychiatry, 46*, 1656–1664.

Ng, S. K. C., Brust, J. C. M., Hauser, W. A., & Susser, M. (1990). Illicit drug use and the risk of new-onset seizures. *American Journal of Epidemiology, 132*, 47–57.

Normand, J., Lempert, R. O., & O'Brien, C. (1994). *Under the influence? Drugs and the American workforce*. Washington, DC: National Academy.

Normand, J. S., Salyards, S., & Mahoney, J. (1990). An evaluation of preemployment drug testing. *Journal of Applied Psychology, 75*, 629–639.

NORML v. Bell, 488 F. Supp. 123 (D.D.C. 1980).

Nowinski, J. (1996). Facilitation 12-step recovery from substance abuse and addiction. In F. Rotgers, D. S. Keller, & J. Morganstern (Eds.), *Treating substance abuse: Theory and technique* (pp. 13–37). New York: Guilford.

Nowinski, J., & Baker, S. (1992). *The twelve-step facilitation handbook*. New York: Lexington Books.

Noyes, R., Brunk, S. F., Avery, D. H., & Canter, A. (1975). The analgesic properties of delta-9-tetrahydrocannabinol and codeine. *Clinical Pharmacology and Therapeutics, 18,* 84–89.

Noyes, R., Brunk, S. F., Baram, D. A., & Canter, A. (1975). Analgesic effects of delta-9-tetrahydrocannabinol. *Journal of Clinical Pharmacology, 15,* 139–143.

Office of the National Drug Control Policy. (1997a). *National drug control strategy.* Washington, DC: Author.

Office of the National Drug Control Policy. (1997b). *State and local spending on drug control activities, Report from the National Survey on local and state governments.* Washington, DC: Author.

Ohlsson, A., Agurell, S., Lindgren, J-E., Gillespie, H. K., & Hollister, L. E. (1985). Pharmacokinetic studies of delta-1-tetrahydrocannabinol in man. In G. Barnett & C. N. Chiang (Eds.), *Pharmacokinetics and pharmacodynamics of psychoactive drugs* (pp. 824–840). Foster City, CA: Biomedical Publications.

Ohlsson, A., Lindgren, J-E., Wahlen, A., Agurell, S., Hollister, L. E., & Gillespie, H. K. (1980). Plasma delta-9-tetrahydrocannabinol concentrations and clinical effects after oral and intravenous administration and smoking. *Clinical Pharmacology and Therapeutics, 28,* 409–416.

Ohlsson, A., Lindgren, J-E., Wahlen, A., Agurell, S., Hollister, L. E., & Gillespie, H. K. (1982). Single-dose kinetics of deuterium-labelled delta-1-tetrahydrocannabinol in heavy and light cannabis users. *Biomedical Mass Spectrometry, 9,* 6–10.

Olsen v. DEA, 878 F.2d 1458 (D.C.C. 1989).

Olsen v. DEA, No. 96–1058, 519 U.S. 1118; 117 S. Ct. 964 U.S. LEXIS 837; 136 L. Ed. 2d 849 U.S.L.W. 3569.

O'Shaughnessy, W. B. (1842). On the preparation of the Indian hemp or gunjah (*Cannabis Indica*): The effects on the animal system in health, and their utility in the treatment of tetanus and other convulsive diseases. *Transactions of the Medical and Physical Society of Bombay, 8,* 421–461.

Ostrowski, J. (1998). Drug prohibition muddles along: How a failure of persuasion has left us with a failed policy (pp. 352–368). In J. M. Fish (Ed.), *How to legalize drugs.* Northvale, NJ: Jason Aronson.

Overholser, J. C. (1987). Clinical utility of the Socratic method. In C. Stout (Ed.), *Annals of clinical research* (pp. 1–7). Des Plaines, IL: Forest Institute.

Pacheco, M. A., Ward, S. J., & Childers, S. R. (1993). Identification of cannabinoid receptors in cultures of rat cerebellar granule cells. *Brain Research, 603,* 102–110.

Packer, H. L. (1968). *The limits of criminal sanction*. Palo Alto, CA: Stanford University Press.

Page, B. J. (1983). The amotivational syndrome hypothesis and the Costa Rica study: Relationships between methods and results. *Journal of Psychoactive Drugs, 15,* 261–267.

Page, B. J., Fletcher, J. M., & True, W. R. (1988). Psychosociocultural perspectives on chronic cannabis use: The Costa Rican follow-up. *Journal of Psychoactive Drugs, 20,* 57–65.

Parish, D. (1989). Relation of pre-employment drug testing result to employment status: A one-year follow-up. *Journal of General Internal Medicine, 4,* 44–47.

Parker, C. S., & Wrigley, F. W. (1950). Synthetic cannabis preparations in psychiatry: I. Synhexyl. *Journal of Mental Science, 96,* 276–279.

Patrick, G., Straumanis, J. J., Struve, F. A., Fitz-Gerald, M. J., Leavitt, J., & Manno, J. E. (1999). Reduced P50 auditory gating response in psychiatrically normal chronic marijuana users: A pilot study. *Biological Psychiatry, 45,* 1307–1312.

Pearl, J., Domino, E., & Rennick, P. (1973). Short-term effects of marijuana smoking on cognitive behavior in experienced male users. *Psychopharmacologia, 31,* 13–24.

Peeke, S. C., Jones, R. T., & Stone, G. C. (1976). Effects of practice on marijuana-induced changes in reaction time. *Psychopharmacology, 48,* 159–163.

Peele, S., with Brodsky, A. (1975). *Love and addiction*. New York: Taplinger.

Peele, S. (1998). *The meaning of addiction*. San Francisco: Josey Bass.

Peels, S. (1989). *The diseasing of America*. Boston: Houghton Mifflin.

Perez-Reyes, M., Timmons, M. C., Davis, K. H., & Wall, E. M. (1973). A comparison of the pharmacological activity in man of intravenously administered delta-9-tetrahydrocannabinol, cannabinol and cannabidiol. *Experientia, 29,* 1368–1369.

Perkins, H. W., Meilman, P. W., Leichliter, J. S., Cashin, J. R., & Presley, C. A. (1999). Misperceptions of the norms for the frequency of alcohol and other drug use on college campuses. *Journal of American College Health, 47,* 253–258.

Perry, D. (1977). Street drug analysis and drug use trends, Part II, 1969–1976. *PharmChem Newsletter, 6,* 4.

Petro, D. J. (1980). Marijuana as a therapeutic agent for muscle spasm or spasticity. *Psychosomatics, 21,* 81–85.

Petro, D. J. (1997a). Pharmacology and toxicity of cannabis. In M. L. Mathre (Ed.), *Cannabis in medical practice* (pp. 56–66). London: McFarland.

Petro, D. J. (1997b). Seizure disorders. In M. L. Mathre (Ed.), *Cannabis in medical practice* (pp. 112–124). London: McFarland.

Petro, D. J. (1997c). Spasticity and chronic pain. In M. L. Mathre (Ed.), *Cannabis in medical practice* (pp. 112–124). London: McFarland.

Petro, D. J., & Ellenberger, C. (1981). Treatment of human spasticity with delta-9-tetrahydrocannabinol. *Journal of Clinical Pharmacology, 21*, 413S–416S.

Pihl, R. O., & Sigal, H. (1978). Motivation levels and the marihuana high. *Journal of Abnormal Psychology, 87*, 280–285.

Plato. (1999). *Great dialogues of Plato*. (W. H. D. Rouse, Trans.). New York: Mass Market Paperback.

Pliny the Elder. (1999). *The natural history*. (H. Rachham, Trans.). Cambridge: Harvard University Press.

Polen, M. R. (1993). Health care use by frequent marijuana smokers who do not smoke tobacco. *Western Journal of Medicine, 158*, 596–601.

Pond, D. A. (1948). Psychological effects in depressive patients of the marijuana homologue synhexyl. *Journal of Neurology, Neurosurgery, and Psychiatry, 11*, 279.

Pope, H. G., & Yurgelun-Todd, D. (1996). The residual cognitive effects of heavy marijuana use in college students. *Journal of the American Medical Association, 275*, 521–527.

Potency Monitoring Project (PMP). (1974–1996). Quarterly Reports. University of Mississippi: Research Institute of Pharmaceutical Sciences.

Powell, W. (1971). *The anarchist cookbook*. Secaucus, NJ: Barricade.

Powers-Lagac, V. (1991). Values clarification approaches to pre-teen substance-abuse prevention. In B. Forster and J. C. Salloway (Eds.), *Prevention and treatments of alcohol and drug abuse* (pp. 119–140). Lewiston, NY: Edwin Mellen.

Prochaska, J. O., & DiClemente, C. C. (1983). Stages and processes of self-change in smoking: Toward an integrative model of change. *Journal of Consulting and Clinical Psychology, 5*, 390–395.

Prochaska, J. O., Norcross, J. C., & DiClemente, C. C. (1994). *Changing for good*. New York: Avon Books.

Project MATCH Research Group. (1998). Matching patients with alcohol disorders to treatments: Clinical implications from project MATCH. *Journal of Mental Health UK, 7*, 589–602.

Quigley, H. A. (1996). Number of people with glaucoma worldwide. *British Journal of Ophthamology, 80*, 389–393.

Rabelais F. (1991). *Gargantua and Pantagruel*. (B. Raffel, Trans.). New York: W. W. Norton.

Raft, D., Gregg, J., Ghia, J., & Harris, L. (1977). Effects of intravenous tetrahydrocannabinol on experimental and surgical pain: Psychological correlates of the analgesic response. *Clinical Pharmacology and Therapeutics, 21,* 26–33.

Rainone, G. A., Deren, S., Kleinman, P. H., & Wish, E. D. (1987). Heavy marijuana users not in treatment: The continuing search for the "pure" marijuana user. *Journal of Psychoactive Drugs, 19,* 353–359.

Randall, R. C., & O'Leary, A. M. (1998). *Marijuana Rx: The patients' fight for medicinal pot.* New York: Thunder's Mouth.

Raspberry, W. (1996, July 15–21). Prevention and the power of persuasion. *Washington Post National Weekly Edition,* p. 29.

Ratcliffe, D. (1974). Summary of street drug results, 1973. *PharmChem Newsletter, 3,* 3.

Ravin v. State, 537 P.2d 494 (Alaska 1975).

Ray, R., Prabhu, G. G., Mohan, D., Nath, L. M., & Neki, J. S. (1979). Chronic cannabis use and cognitive functions. *Indian Journal of Medical Research, 69,* 996–1000.

Razdan, R. K. (1986). Structure-activity relationships in cannabinoids. *Pharmacology Review, 38,* 75–149.

Reilly, D., Didcott, P., Swift, W., & Hall, W. (1998). Long-term cannabis use: Characteristics of users in an Australian rural area. *Addiction, 93,* 837–846.

Reynolds, J. R. (1890). On the therapeutic uses and toxic effects of *Cannabis Indica. Lancet, 1,* 637–638.

Richter, A., & Loscher, W. (1994). (+)-WIN55,212–2 A novel cannabinoid receptor agonist, exerts antidystonic effects in mutant dystonic hamsters. *European Journal of Pharmacology, 264,* 371–377.

Riedel, W. J., Vermeeren, A., Van Boxtel, M. P. J., Vuurman, E. F. P. M., Verhey, F. R. J., Jolles, J., et al. (1998). Mechanisms of drug-induced driving impairment: A dimensional approach. *Human Psychopharmacology, 13,* S49–S63.

Robbe, H. (1998). Marijuana's impairing effects on driving are moderate when taken alone but severe when combined with alcohol. *Human Psychopharmacology: Clinical and Experimental, 13,* S70–S78.

Robinson, J. (1994). Why Germans get six weeks off and you don't. *Escape,* Winter. Available: http://www.escapemag.com/home/sub_3c.htm.

Rochford, J., Grant, I., & LaVigne, G. (1977). Medical students and drugs: Further neuropsychological and use pattern considerations. *International Journal of the Addictions, 12,* 1057–1065.

Roffman, R. A. (1982). *Marijuana as medicine.* Seattle: Madrona Publishers.

Roffman, R. A., & Barnhart, R. (1987). Assessing need for marijuana dependence treatment through an anonymous telephone interview. *International Journal of the Addictions, 22,* 639–651.

Roffman, R. A., Klepsch, R., Wertz, J. S., Simpson, E. E., & Stephens, R. S. (1993). Predictors of attrition from an outpatient marijuana-dependence counseling program. *Addictive Behaviors, 18,* 553–566.

Roffman, R. A., & Stephens, R. S. (1993). Cannabis dependence. In D. L. Dunner (Ed.), *Current psychiatric therapy* (pp. 105–109). Philadelphia: Saunders.

Rogers, C. (1950). A current formulation of client-centered therapy. *Social Service Review, 24,* 442–450.

Rohrich, J., Zorntlein, S., Potsch, L., Skopp, G., & Becker, J. (2000). Effect of the shampoo Ultra Clean on drug concentrations in human hair. *International Journal of Legal Medicine, 113,* 102–106.

Rosenkrantz, H. (1976). The immune response and marijuana. In G. Nahas, W. D. Paton, and J. Idanpaan-Heikkila (Eds.), *Marihuana: Chemistry, biochemistry and cellular effects* (pp. 441–456). New York: Springer-Verlag.

Rosenthal, E., Gieringer, D., & Mikuriya, T. (1997). *Marijuana medical handbook.* Oakland: Quick American Archives.

Rosenthal, E. & Kubby, S. (1996). *Why marijuana should be legal.* New York: Thunder's Mouth.

Rosenthal, F. (1971) *The herb.* Leiden: E. J. Brill.

Rosenthal, M. S., & Kleber, H. D. (1999). Making sense of medical marijuana. *Proceedings of the Association of American Physicians, 111,* 159–165.

Roth, M. D., Arora, A. Barsky, S. H., Kleerup, E. C., Simmons, M., & Tashkin, D. P. (1998). Airway inflammation in young marijuana and tobacco smokers. *American Journal of Respiratory and Critical Care Medicine, 157,* 928–937.

Roth, M. D., Kleerup, E. C., Arora, A., Barsky, S. H., & Tashkin, D. P. (1996). Endobronchial injury in young tobacco and marijuana smokers as evaluated by visual, pathologic and molecular criteria. *American Journal of Respiratory and Critical Care Medicine, 153,* 100A.

Rosenthal, R., & Rosnow, R. L. (1991). *Essentials of behavioral research.* New York: McGraw-Hill.

Roueche, B. (1963). Alcohol in human culture. In S. P. Lucia (Ed.), *Alcohol and civilization* (pp. 167–182). New York: McGraw-Hill.

Rowell, E. A., & Rowell, R. (1939). *On the trail of marijuana, the weed of madness.* Mountain View, CA: Pacific.

Rubin, V. (Ed). (1975). *Cannabis and culture.* The Hague: Mouton.

Rubin, V., & Comitas, L. (1975). *Ganja in Jamaica, a medical anthropological study of chronic marihuana use.* The Hague: Mouton.

Rudenko, S. I. (1970). *Frozen tombs of Siberia.* Berkeley: University of California Press.

Russell, J. M., Newman, S. C., & Bland, R. C. (1994). Drug abuse and dependence. *Acta Psychiatrica Scandinavica, 376*(Suppl.), 54–62.

Russo, E. (1998). Cannabis for migraine treatment: The once and future prescription? An historical and scientific review. *Pain, 76,* 3–8.

Sallan, S. E., Zinberg, N. E., & Frei, E. (1975). Antiemetic effects of delta-9-tetrahydrocannabinol in patients receiving cancer chemotherapy. *New England Journal of Medicine, 293,* 795–797.

Sandyk, R., & Awerbuch, G. (1988). Marijuana and Tourette's syndrome. *Journal of Clinical Psychopharmacology, 8,* 444–445.

Sanudo-Pena, M. C., & Walker, J. M. (1997). Role of subthalamic nucleus in cannabinoid action in the substantia nigra of the rat. *Journal of Neurophysiology, 77,* 1635–1638.

Sanudo-Pena, M. C., & Walker, J. M. (1998). Effects of intrastitial cannabinoids on rotational behavior in rats: Interactions with the dopaminergic system. *Synapse, 30,* 221–226.

Satz, P., Fletcher, J. M., & Sutker, L. S. (1976). Neuropsychologic, intellectual and personality correlates of chronic marijuana use in native Costa Ricans. *Annals of the New York Academy of Sciences, 282,* 266–306.

Schaeffer, J., Andrysiak, T., & Ungerleider, J. T. (1981). Cognition and long-term use of Ganja (cannabis). *Science, 213,* 465–466.

Schenk, S., & Partridge, B. (1999). Cocaine-seeking produced by experimenter-administered drug injections: Dose-effect relationships in rats. *Psychopharmacology, 147,* 285–290.

Schinke, S. P., Botvin, G. J., & Orlandi, M. A. (1991). *Substance abuse in children and adolsecents: Evaluation and intervention.* Newbury Park, CA: Sage.

Schmitz, J. M., Oswald, L. M., Jacks, S. D., Rustin, T., Rhoades, H. M., & Grabowski, J. (1997). Relapse prevention treatment for cocaine dependence: Group vs. individual format. *Addictive Behaviors, 22,* 405–418.

Schneider, A., & Flaherty, M. P. (1991, August 11). Presumed guilty: The law's victims in the war on drugs. *The Pittsburgh Press.* Available: http://www.taima.org/wod/wod.htm.

Schneier, F. R., & Siris, S. G. (1987). A review of psychoactive substance use and abuse in schizophrenia: Patterns of drug choice. *Journal of Nervous and Mental Disease, 175,* 641–652.

Schuckit, M. A., Daeppen, J. B., Danko, G. P., Tripp, M. L., Smith, T. L., Li, T. K., et al. (1999). Clinical implications for four drugs of the DSM-IV distinction between substance dependence with and without a physiological component. *American Journal of Psychiatry, 156,* 41–49.

Schuel, H., Chang, M. C., Burkman, L. J., Picone, R. P., Makriyannis, A., Zimmerman, A. M., et al. (1999). Cannabinoid receptors in sperm. In G. G. Nahas, K. M. Sutin, D. J. Harvey, & S. Agurell (Eds.), *Marijuana and medicine* (pp. 335–346). Totowa, NJ: Humana.

Schultes, R. E., Klein, W. M., Plowman, T., & Lockwood, T. E. (1975). Cannabis: An example of taxonomic neglect. In V. Rubin (Ed.), *Cannabis and culture* (pp. 21–38). The Hague: Mouton.

Schwartz, R. H. (1984). Marijuana: A crude drug with a spectrum of unappreciated toxicity. *Pediatrics, 73,* 457.

Schwartz, R. H. (1991). Heavy marijuana use and recent memory impairment. *Psychiatric Annals, 21,* 80–82.

Schwartz, R. H., Gruenewald, P. J., Klitzner, M., & Fedio, P. (1989). Short-term memory impairment in cannabis-dependent adolescents. *American Journal of Diseases of Children, 143,* 1214–1219.

Scott, J. M. (1969). *The white poppy: A history of opium.* New York: Funk and Wagnalls.

Shahar, A., & Bino, T. (1974). In vitro effects of delta-9-tetrahydrocannabinol (THC) on bull sperm. *Biochemical Pharmacology, 23,* 1341–1342.

Sharma, S., & Moskowitz, H. (1974). Effects of two levels of attention demand on vigilance performance under marihuana. *Perceptual and Motor Skills, 38,* 967–970.

Shedler, J., & Block, J. (1990). Adolescent drug use and psychological health: A longitudinal inquiry. *American Psychologist, 45,* 612–630.

Shen, M., Piser, T. M., Seybold, V. S., & Thayer, S. A. (1996). Cannabinoid receptor agonists inhibit glutamatergic synaptic transmission in rat hippocampal cultures. *Journal of Neuroscience, 16,* 4322–4334.

Shope, J. T., Copeland, L. A., Kamp, M. E., & Lang, S. W. (1998). Twelfth grade follow-up of the effectiveness of a middle school-based substance abuse prevention program. *Journal of Drug Education, 28,* 185–197.

Sidney, S., Quesenberry, C. P., Friedman, G. D., & Tekawa, I. S. (1997). Marijuana use and cancer incidence (California, United States). *Cancer Cause and Control, 8,* 722–728.

Simon, T. R., Stacy, A. W., Sussman, S., & Dent, C. W. (1994). Sensation seeking and drug use among high risk Latino and Anglo adolescents. *Personality and Individual Differences, 17,* 665–672.

Simonds, J. F., & Kashani, J. (1980). Specific drug use and violence in delinquent boys. *American Journal of Drug and Alcohol Abuse, 7,* 305–322.

Simons, J., Correia, C. J., Carey, K. B., & Borsari, B. E. (1998). Validating a five-factor marijuana motives measure: Relations with use, problems, and alcohol motives. *Journal of Counseling Psychology, 45,* 265–273.

Slikker, W., Paule, M. G., Ali, S. F., Scallett, A. C., & Bailey, J. R. (1992). Behavioral, neurochemical, and neurohistological effects of chronic marijuana smoke exposure in the nonhuman primate. In L. Murphy & A. Bartke (Eds.), *Marijuana/cannabinoids: Neurobiology and neurophysiology* (pp. 387–423). Boca Raton, FL: CRC.

Sloman, L. (1998). *Reefer madness: A history of marijuana.* New York: St. Martin's Griffin.

Smiley, A. (1986). Marijuana: On-road and driving simulator studies. *Alcohol, Drugs, and Driving, 2,* 121–134.

Smith, C. G., Almirez, R. G., Scher, P. M., & Asch, R. H. (1984). Tolerance to the reproductive effects of delta-9-tetrahydrocannabinol. In S. Agurell, W. Dewy, and R. Willette (Eds.), *The cannabinoids: Chemical, pharmacologic, and therapeutic aspects* (pp. 471–485). New York: Academic.

Smith, D. E. (1968). The acute and chronic toxicity of marijuana. *Journal of Psychedelic Drugs, 2,* 37–48.

Smith, F. L., Fujimori, K., Lowe, J., & Welch, S. P. (1998). Characterization of delta-9-tetrahydrocannabinol and anandamide antinociception in nonarthritic and arthritic rats. *Pharmacology, Biochemistry and Behavior, 60,* 183–191.

Sobell, L. C., Sobell, M. B., Cunningham, J. A., & Toneatto, T. (1993). A life-span perspective on natural recovery (self-change) from alcohol problems. In J. S. Baer, G. A. Marlatt, & R. J. McMahon (Eds.), *Addictive behaviors across the life span* (pp. 34–68). Newbury Park, CA: Sage.

Solomon, D. (1966). *The marijuana papers.* New York: Bobbs-Merrill.

Solowij, N. (1998). *Cannabis and cognitive functioning.* Cambridge: Cambridge University Press.

Soueif, M. I. (1976). Some determinants of psychological deficits associated with chronic cannabis consumption. *Bulletin on Narcotics, 28,* 25–42.

Spunt, B., Goldstein, P., Brownstein, H., & Fendrich, M. (1994). The role of marijuana in homicide. *International Journal of the Addictions, 29,* 195–213.

Staquet, M., Gantt, C., & Machlin, D. (1978). Effect of nitrogen analog of tetrahydrocannabinol on cancer pain. *Clinical Pharmacology and Therapeutics, 23,* 397–401.

Steele, N., Gralla, R. J., & Braun, D. W. (1980). Double-blind comparison of antiemetic effects of nabilone and prochlorperazine on chemotherapy-induced emesis. *Cancer Treatment Reports, 64,* 219–224.

Stefanis, C. (1976). Biological aspects of cannabis use. In R. C. Petersen (Ed.), *The international challenge of drug abuse* (pp. 149–178). Rockville, MD: National Institute of Drug Abuse.

Stefanis, C., Ballas, C., & Madianou, D. (1975). Sociocultural and epidemiological aspects of hashish use in Greece. In V. Rubin (Ed.), *Cannabis and culture* (pp. 303–326). The Hague: Mouton.

Stefano, G., Salzet, B., & Salzet, M. (1997). Identification and characterization of the leech CNS cannabinoid receptor: Coupling to nitric oxide release. *Brain Research, 753,* 219–224.

Stein, J. A., Newcomb, M. D., & Bentler, P. M. (1996). Initiation and maintenance of tobacco smoking: Changing determinants and correlates in adolescence and young adulthood. *Journal of Applied Social Psychology, 26,* 160–187.

Stella, N., Schweitzer, P., & Piomelli, D. (1997). A second endogenous cannabinoid that modulates long-term potentiation. *Nature, 388,* 773–778.

Stephens, R. S., Curtin, L., Simpson, E. E., & Roffman, R. A. (1994). Testing the abstinence violation effect construct with marijuana cessation. *Addictive Behaviors, 19,* 23–32.

Stephens, R. S., Roffman, R. A., & Simpson, E. E. (1993). Adult marijuana users seeking treatment. *Journal of Consulting and Clinical Psychology, 61,* 1100–1104.

Stephens, R. S., Roffman, R. A., & Simpson, E. E. (1994). Treating adult marijuana dependence: A test of the relapse prevention model. *Journal of Consulting and Clinical Psychology, 62,* 92–99.

Stiglick, A., & Kalant, H. (1982a). Residual effects of prolonged cannabis administration on exploration and DRL performance in rats. *Psychopharmacology, 77,* 124–128.

Stiglick, A., & Kalant, H. (1982b). Learning impairment in the radial-arm maze following prolonged cannabis treatment in rats. *Psychopharmacology, 77,* 117–23.

Stockings, G. T. (1947). A new euphoriant for depressive mental states. *British Medical Journal, 1,* 918–922.

Stoltenberg, J. (1988). *Refusing to be a man: Essays on sex and justice.* New York: Meridian.

Strohmetz, D. B., Alterman, A. I., & Walter, D. (1990). Subject selection bias in alcoholics volunteering for a treatment study. *Alcoholism: Clinical and Experimental Research, 14,* 736–738.

Strupp, H. H. (1989). Psychotherapy: Can the practitioner learn from the researcher? *American Psychologist, 44,* 717–724.

Seruve, F., Straurmanis, J. J., Patrick, G., Leavitt, J., Manno, J. E., & Manno, B. R. (1999). Topographic quantitative EEG sequelae of chronic marijuana users. *Drug and Alcohol Dependence, 56,* 167–179.

Substance Abuse and Mental Health Services Administration (SAMHSA). (1997). *National Household Survey on Drug Abuse: Population estimates, 1996.* Rockville, MD: U.S. Department of Health and Human Services.

Substance Abuse and Mental Health Services Administration (SAMHSA). (2000). *Summary of findings from the 1999 National Household Survey on Drug Abuse*. Rockville, MD: SAMHSA.

Sugiura, T., Kodaka, T., Nakane, S., Miyashita, T., Kondo, S., Suhara, Y., et al. (1999). Evidence that the cannabinoid CB1 receptor is a 2-arachidonoylglycerol receptor. Structure-activity relationship of 2-arachidonoylglycerol, ether-linked analogues, and related compounds. *Journal of Biological Chemistry, 274,* 2794–2781.

Sugiura, T., Kondo, S., Kishimoto, S., Miyashita, T., Nakan, S., Kodaka, T., et al. (2000). Evidence that 2-arachidonoylglycerol but no N-palmitoylethanolamine or anadamide is the physiological ligand for the cannabinoid CB2 receptor. Comparison of the agonistic activities of various cannabinoid receptor ligands in HL-60 cells. *Journal of Biological Chemistry, 275,* 605–612.

Sussman, S., Dent, C. W., Stacy, A. W., & Craig, S. (1998). One-year outcomes of Project Towards No Drug Abuse. *Preventive Medicine, 27,* 632–642.

Sussman, S., Simon, T. R., Dent, C. W., Steinberg, J. M., & Stacy, A. W. (1999). One-year prediction of violence perpetration among high-risk youth. *American Journal of Health Behavior, 23,* 332–344.

Sussman, S., Stacy, A. W., Dent, C. W., Simon, T. R., & Johnson, C. A. (1996). Marijuana use: Current issues and new research directions. *Journal of Drug Issues, 26,* 695–733.

Sutherland, G., Stapleton, J. A., Russell, M. A. H., Jarvis, J. J., Hajek, P., Belcher, M., et al. (1992). Randomized controlled trial of nasal nicotine spray in smoking cessation. *Lancet, 340,* 324–329.

Swan, N. (1994). A look at marijuana's harmful effects. *NIDA Notes, 9,* 17–42.

Szasz, T. S. (1961). *The myth of mental illness*. New York: Hoeber-Harper.

Szasz, T. (1992). *Our right to drugs*. New York: Praeger.

Tart, C. T. (1971). *On being stoned*. Palo Alto, CA: Science and Behavior Books.

Tashkin, D. P. (1999). Marijuana and the lung. In G. G. Nahas, K. M. Sutin, D. J. Harvey, & S. Agurell (Eds.), *Marijuana and medicine* (pp. 279–288). Totowa, NJ: Humana.

Tashkin, D. P., Coulson, A. H., Clark, V. A., Simmons, M., Bourque, L. B., Duann, S., et al. (1987). Respiratory symptoms and lung function in habitual, heavy smokers of marijuana alone, smokers of marijuana and tobacco, smokers of tobacco alone, and nonsmokers. *American Review of Respiratory Disease, 135,* 209–216.

Tashkin, D. P., Simmons, M. S., Sherrill, D. L., & Coulson, A. H. (1997). Heavy habitual marijuana smoking does not cause an accelerated decline in FEV1 with age. *American Journal of Respiratory and Critical Care Medicine, 155,* 141–148.

Taylor, B. (1854). *A journey to central Africa.* New York: Putnam.

Taylor, B. (1855). *The land of the Saracens; or pictures of Palestine, Asia Minor, Sicily and Spain.* New York: Putnam.

Taylor, S., Vardaris, R., Rawitch, A., Gammon, C., Cranston, J., & Lubetkin, A. (1976). The effects of marijuana on human physical aggression. *Aggressive Behavior, 2,* 153–161.

Terhune, K. W., Ippolito, C. A., & Crouch, D. J. (1992). *The incidence and role of drugs in fatally injured drivers* (DOT HS Report No. 808 065). Washington DC: U.S. Department of Transportation, National Highway Traffic Safety Administration.

Thistle, J., & Cook, J. P. (1972). *Seventeenth-century economic documents.* Oxford: Clarendon Press.

Thornicroft, G. (1990). Cannabis and psychosis: Is there epidemiological evidence for an association? *British Journal of Psychiatry, 157,* 25–33.

Timpone, J. G., Wright, D. J., Li, N., Egorin, M. J., Enama, M. E., Mayers, J., et al. (1997). The safety and pharmacokinetics of single-agent and combination therapy with megesterol acetate and dronabinol for the treatment of HIV wasting syndrome. *AIDS Research and Human Retroviruses, 13,* 305–315.

Tinklenberg, J. R., Murphy, P., Murphy, P. L., & Pfefferbaum, A. (1981). Drugs and criminal assaults by adolescents: A replication study. *Journal of Psychoactive Drugs, 13,* 277–287.

Trent, L. K. (1998). Evaluation of a four- versus six-week length of stay in the Navy's alcohol treatment program. *Journal of Studies on Alcohol, 59,* 270–279.

Truong, X. T., & Hanigan, W. C. (1986). Effect of delta-9-THC on EMG measurements in human spasticity. *Clinical Pharmacology and Therapeutics, 39,* 232.

Tucker, J. A., Donovan, D. M., & Marlatt, G. A. (Eds.). (1999). *Changing addictive behavior: Bridging clinical and public health strategies.* New York: Guilford.

Tunving, K., Thulin, O., Risberg, J., & Warkentin, S. (1986). Regional cerebral blood flow in long-term heavy cannabis use. *Psychiatry Research 17,* 15–21.

Turner, C. E., & ElSohly, M. A. (1981). Biological activity of cannabichromene, its homologs and isomers. *Journal of Clinical Pharmacology, 21,* 283S–291S.

Turner, C. E., & Hadley, K. W. (1974). Chemical analysis of cannabis sativa of distinct origin. *Archivos de Investigacion Medica, 5*, 141–150.

Tusser, T. (1580). *Five hundred points of good husbandrie*. London: Henrie Denham.

Tyson, L. B., Gralla, R. J., Clark, R. A., Kris, M. G., Bordin, L. A., & Bosl, G. J. (1985). Phase 1 trial of levonantradol in chemotherapy-induced emesis. *American Journal of Clinical Oncology, 8*, 528–532.

Uestuen, B., Compton, W., Mager, D., Babor, T., Baiyewu, O., Chatterji, S., et al. (1997). WHO study on the reliability and validity of the alcohol and drug use disorder instruments: Overview of methods and results. *Drug and Alcohol Dependence, 47*, 161–169.

Urquhart, D. (1855). *The pillars of Hercules; or a narrative of travels in Spain and Morocco in 1848*. New York: Harper.

Vachon, L., Sulkowski, A., & Rich, E. (1974). Marihuana effects on learning, attention and time estimation. *Psychopharmacologia, 39*, 1–11.

Vaillant, G. E. (1983). *The natural history of alcoholism*. Cambridge: Harvard University Press.

Van Tulder, M. W., Cherkin, D. C., Berman, B., Lao, L., & Koes, B. W. (1999). The effectiveness of acupuncture in the management of acute and chronic low back pain. *Spine, 24*, 1113–1123.

Van der Merwe, N. J. (1975). Cannabis smoking in 13th–14th century Ethiopia. In V. Rubin (Ed.), *Cannabis and culture* (pp. 77–80). The Hague: Mouton.

Vinciguerra, V., Moore, T., & Brennan, E. (1988). Inhalation marijuana as an antiemetic for cancer chemotherapy. *New York State Journal of Medicine, 88*, 525–527.

Volicer, L., Stelly, M., Morris, J., McLaughlin, J., & Volicer, B. J. (1997). Effects of dronabinol on anorexia and disturbed behavior in patients with Alzheimer's disease. *International Journal of Geriatric Psychiatry, 12*, 913–919.

Volkow, N. D., Gillespie, H., Mullani, N., Tancredi, L., Grant, C., Valentine, A., et al. (1996). Brain glucose-metabolism in chronic marijuana users at baseline and during marijuana intoxication. *Psychiatry Research: Neuroimaging, 67*, 29–38.

Volpicelli, J. R., Alterman, A. I., Hayashida, M., & O'Brien, C. P. (1992). Naltrexone in the treatment of alcohol dependence. *Archives of General Psychiatry, 49*, 876–880.

Von Bibra, E. (1855/1994). *The narcotic luxury: Hemp and humans*. Lohrbach: Werner Piper's Medien Xperimente.

Wall, M. E., Sadler, B. M., Brine, D., Harold, T., & Perez-Reyes, M. (1983). Metabolism, disposition, and kinetics of delta-9-tetrahydrocannabinol in men and women. *Clinical Pharmacology and Therapeutics, 34,* 352–363.

Wallace, J. (1990). Controlled drinking, treatment effectiveness, and the disease model of addiction: A commentary on the ideological wishes of Stanton Peele. *Journal of Psychoactive Drugs, 22,* 261–284.

Wallace, J. (1996). Theory of 12-step oriented treatment. In F. Rotgers, D. S. Keller, & J. Morganstern (Eds.), *Treating substance abuse: Theory and technique* (pp. 13–37). New York: Guilford.

Wallnofer, H., & Von Rottauscher, A. (1965). *Chinese folk medicine and acupuncture.* New York: Bell.

Wampold, B. E., Mondin, G. W., Moody, M., Stich, F., Benson, K., & Ahn, H. (1997). A meta-analysis of outcome studies comparing bona fide therapies: Empirically, "all must have prizes." *Psychological Bulletin, 122,* 203–215.

Weckowicz, T. E., Collier, G., & Spreng, L. (1977). Field dependence, cognitive functions, personality traits, and social values in heavy cannabis users and nonuser controls. *Psychological Reports, 41,* 291–302.

Weisheit, R. A. (1992). *Domestic marijuana: A neglected industry.* New York: Greenwood.

Weller, R. A., & Halikas, J. A. (1980). Objective criteria for the diagnosis of marijuana abuse. *Journal of Nervous and Mental Disease, 176,* 719–725.

Weller, R. A., & Halikas, J. A. (1982). Change in effects from marijuana: A five- to six-year follow-up. *Journal of Clinical Psychiatry, 43,* 362–365.

Weller, R. A., & Halikas, J. A. (1984). Marijuana use and sexual behavior. *Journal of Sex Research, 20,* 186–193.

Welte, J. W., & Barnes, G. M. (1985). Alcohol: The gateway to other drug use among secondary-school students. *Journal of Youth and Adolescence, 14,* 487–498.

Werner, J. (1964). Frankish royal tombs in the cathedrals of Cologne and Saint Denis. *Antiquity, 38,* 201–216.

West, M. (1997). The use of certain cannabis derivatives (Canasol) in glaucoma. In M. L. Mathre (Ed.), *Cannabis in medical practice* (pp. 103–111). London: McFarland.

Wetzel, C. D., Janowsky, D. S., & Clopton, P. L. (1982). Remote memory during marijuana intoxication. *Psychopharmacology, 76,* 278–281.

White, H. R., & Hansell, S. (1998). Acute and long-term effects of drug use on aggression from adolescence into adulthood. *Journal of Drug Issues, 28,* 837–858.

White, H. R., Loeber, R., Stouthamer-Loeber, M., & Farrington, D. (1999). Developmental associations between substance use and violence. *Development and Psychopathology, 11,* 785–803.

Whittier, J. G. (1854/1904). *The Compleat Poetical Works of John Greenleaf Whittier*. Boston: Houghton Mifflin.

Wig, N. N., & Varma, V. K. (1977). Patterns of long-term heavy cannabis use in north India and its effects on cognitive functions: A preliminary report. *Drug and Alcohol Dependence, 2,* 211–219.

Wiley, J. L. (1999). Cannabis: Discrimination of "internal bliss"? *Pharmacology, Biochemistry and Behavior, 64,* 257–260.

Williams, A. F., Peat, M. A., & Crouch, D. J. (1985). Drugs in fatally injured young male drivers. *Public Health Reports, 100,* 19–25.

Williams, C. M., & Kirkham, T. C. (1999). Anandamide induces overeating: Mediation by central cannabinoid (CB1) receptors. *Psychopharmacology, 143,* 315–317.

Wilson, W., Mathew, R., Turkington, T., Hawk, T., Coleman, R. E., & Provenzale, J. (2000). Brain morphological changes and early marijuana use: A magnetic resonance and positron emission tomography study. *Journal of Addictive Diseases, 19,* 1–22.

Wimbush, J. C., & Dalton, D. R. (1997). Base rate of employee theft: Convergence of multiple methods. *Journal of Applied Psychology, 82,* 756–763.

Wirtshafter, D. (1997). Nutritional value of hemp seed and hemp seed oil. In M. L. Mathre (Ed.), *Cannabis in medical practice* (pp. 181–191). London: McFarland.

Witter, F. R., & Niebyl, J. R. (1990). Marijuana use in pregnancy and pregnancy outcome. *American Journal of Perinatology, 7,* 36–38.

Wood, G. B., & Bache, F. (1868). *The dispensatory of the United States of America* (18th ed.) (pp. 379–382). Philadelpia: Lippincott.

Woody, G. E., & MacFadden, W. (1995). Cannabis related disorders. In H. I. Kaplan & B. J. Sadock (Eds.), *Comprehensive textbook of psychiatry* (6th ed.) (pp. 810–817). Baltimore: Williams & Wilkins.

Yonkers, K. A., Warshaw, M. G., Massion, A. O., & Keller, M. B. (1996). Phenomenology and course of generalised anxiety disorder. *British Journal of Psychiatry, 168,* 308–313.

Yoshida, H., Usami, N., Ohishi, Y., Watanabe, K., Yamamoto, I., & Yoshimura, H. (1995). Synthesis and pharmacological effects in mice of halogenated cannabinol derivatives. *Chemical and Pharmaceutical Bulletin, 42,* 335–337.

Yuille, J. C., Tollestrup, P. A., Marxsen, D., Porter, S., & Herve-Hugues, F. M. (1998). An exploration on the effects of marijuana on eyewitness memory. *International Journal of Law & Psychiatry, 21,* 117–128.

Zacny, J. P., & Chait, L. D. (1989). Breathhold duration and response to marijuana smoke. *Pharmacology, Biochemistry and Behavior, 33,* 481–484.

Zacny, J. P., & Chait, L. D. (1991). Response to marijuana as a function of potency and breathhold duration. *Psychopharmacology, 103,* 223–226.

Zeese, K. (1997). Legal issues related to the medical use of marijuana. In M. L. Mathre (Ed.), *Cannabis in medical practice* (pp. 20–32). London: McFarland.

Zias, J., Stark, H., Seligman, J., Levy, R., Werker, E., Breur, A., et al. (1993). Early medical use of cannabis. *Nature, 363,* 215.

Zimmer, L., & Morgan, J. P. (1997). *Marijuana myths marijuana facts.* New York: The Lindesmith Center.

Zimmerman, A. M., Zimmerman, S., & Raj, A. Y. (1979). Effects of cannabinoids on spermatogenesis in mice. In G. G. Nahas and W. D. M. Paton (Eds.), *Marijuana: Biological effects, analysis, metabolism, cellular responses, reproduction, and brain* (pp. 407–418). New York: Pergamon.

Zinberg, N. E. (1984). *Drug set and setting: The basis for controlled intoxicant use.* New Haven: Yale University Press.

Zuardi, A. W., Cosme, R. A., Graeff, F. G., & Guimaraes, F. S. (1993). Effects of ipsapirone and cannabidiol on human experimental anxiety. *Journal of Psychopharmacology, 7,* 82–88.

Zuardi, A. W. & Guimaraes, F. S. (1997). Cannabidiol as an anxiolytic and antipsychotic. In M. L. Mathre (Ed.), *Cannabis in medical practice* (pp. 133–141). London: McFarland.

Zwerling, C., Ryan, J., & Orav, E. J. (1990). The efficacy of preemployment drug screening for marijuana and cocaine in predicting employment outcomes. *Journal of the American Medical Association, 264,* 2639–2643.

Index